GW01326531

Buddhist Revivalist Movements

Alan Robert Lopez

Buddhist Revivalist Movements

Comparing Zen Buddhism and the Thai Forest Movement

Alan Robert Lopez
Chiang Mai, Thailand

ISBN 978-1-137-54349-3 ISBN 978-1-137-54086-7 (eBook)
DOI 10.1057/978-1-137-54086-7

Library of Congress Control Number: 2016956808

Printed on acid-free paper

This Palgrave Macmillan imprint is published by Springer Nature
The registered company is Nature America Inc. New York

Preface

This story begins 2500 years ago in Northeast India in the foothills of the Himalayas. Here were the original forest monks of the Buddha's newly founded sangha. Here also is the origin of the doctrines and practices that would metamorphose and travel eastward to become Ch'an Buddhism, better known in the West under its Japanese name Zen. A more proximate starting point is several decades ago in the foothills of the Berkshires in Northwestern Connecticut. Here in a retreat that attracted Buddhists from different denominations; participants practiced a meditation that claimed to point directly at the essence of mind. We listened to talks from a cross section of Buddhist traditions. In this ecumenical space grounded in practice, a message that perhaps had been intuited by many of us was clearly enunciated. Each day, different dharma teaching was invited to speak. Among them was a monk of the Thai Forest Tradition, a disciple of the famed Ajahn Chah. What he told us to his delight and wonderment was that hearing the teachings on the nature of mind and the practices that accompanied them was like listening to his Thai teacher. He assured us that no monk had come over the mountains north of Thailand and descended to his forest wat or monastery. The practice/wisdom of his Thai wat was homegrown.

I had been handed a *genjo koan*, a puzzle from everyday life. How could schools of different cultures, history, geography, and most significantly of different philosophies evidence such similarities? Of course, there is always the romantic, universalist argument that all is one and all teachings are really the same, and all roads lead home. Yet, such a gloss was not a satisfying answer to my koan.

At the time, I had little knowledge of the Thai Forest Tradition other than a respect for the clean lines of Theravada thought and training. The majority of my practice had been Zen, but more recently, I had begun engaging a Tibetan teaching which had an affinity with Zen. Trips to Nepal for retreats and trekking also brought me to stopovers in Thailand. Stopovers led to an extended relocation and the opportunity for considerable contact with Thai Buddhism and especially the Forest Tradition.

A Comparative Study

The question that had been framed was, of course, inherently comparative. Yet like any question that is held it evolved. The evolved question demanded a closer investigation and in turn spawned additional and more pointed queries. What remained was the comparative framework that required going beyond a single ethnography. The more specific questions that arose and will be addressed in this study ask in what ways the two schools, Ch'an/Zen and the Forest Tradition, are similar (and different). What characteristics of these movements seem to generate or underpin their affinities? What factors engender comparable styles, practices, and even doctrines in groups that have had no direct communication over the centuries? As their affinities cannot be deduced by their designations as Theravada or Mahayana, what categories may be effective in isolating and explaining the characteristics of the mountain sect, Ch'an/Zen, and the forest sect?

Obviously, a comparative strategy entails a cognitive posture that can be questioned. To compare means, to some extent, to de-contextualize. The reference point is no longer the unique socio-historical setting but is the other case. The settings of Ch'an/Zen and the Forest schools remain relevant but as points of distinction or resemblance to one another. The study of both these sects has been advanced in recent years by delving deeper into their cultural milieu. Stanley Tambiah in his examination of the monks of Northeast Thailand speaks of "total culture" (1976). The archeology of Ch'an has unearthed unexpected dimensions of the self-proclaimed meditation sect (Faure 2003). A comparative investigation seems to move in the opposite direction. Rather than greater contextualizing and historical specifying, a comparison identifies general outline and lifts to some extent each phenomenon out of its cultural context.

A comparative approach permits inquiries that would be ignored if one is restricted to a strongly contextualized case study. First, it is important to acknowledge that all knowledge is comparative. To imagine a reference-less description is to conjure a near useless and unintelligible communication. With comparison factors at play can be perceived and conceptualized that otherwise might go unnoticed. Analytical categories and typologies can be developed that offer more than intricate description. De-contextualization has its risks but also its rewards.

A corollary to the comparative stance and one that we will return to is the limits of approaching these sects through the denominational and formal doctrinal labels: Theravada and Mahayana. If we take these labels as primary, why even search out affinities? Why compare apples and oranges? What our approach suggests is that categories like Theravada/Mahayana and Thai/Chinese may at times occlude more than they reveal.

A story told by Jack Kornfield is relevant. The Karmapa, the head of one of the main Tibetan schools, was invited to speak at a meditation center during his American tour. Translation was provided for a Thai monk who accompanied Jack. Midway through the talk, he exclaimed, "He's a Buddhist!" Sometimes, we need to listen past the alien tongue and look past the exotic garment to recognize mutuality. This study will seek to do just that.

Chiang Mai, Thailand — Alan Robert Lopez

Acknowledgments

Buddhism tells us that all arisings are connected to a web of causes and conditions. This book is no exception. Acknowledgment of those who constitute the web that brought this work into existence is called for. Conditions are both distant and proximate. Acknowledgment of near support must begin with Dr. Brooke Schedneck whose invitation to present on this theme resulted in a paper, now a chapter. That a paper became a book is due to the interest and opportunity offered by Palgrave Macmillan. Their continued guidance during the course of the writing deserves thanks.

Of course, gratitude is offered to my teachers, secular and spiritual, human and non-human, too many to mention. Thank you for not leaving for Nibbana and hanging around to teach me more than I will ever know.

Giving and receiving are ubiquitous. A Zen Master called this "giving life to life." My deepest gratitude goes to those who have been willing to share their life with mine: to my daughter, Jasmine, who kept joy alive during the dry and difficult moments; to my wife, Noi, whose unstinting efforts kept our domestic life on course during my solitary labors; to my friends in the Thursday men's circle for providing me a forum for my frustrations and for reminding me what it's all about; and to my long-distance comrades via computer calls who reminded me that physical distance need not reduce empathy or affection.

And finally to my parents, Bob and Wanda, who for all their struggles and human limitations manage to transmit to me a confidence in my endeavors.

Alan Robert Lopez

Contents

CHAPTER 1

Entering the Mountains and Forests

Introduction

The mountains of East Asia and the forests of South Asia are ecological contrasts that share a social affinity. Their differences have presented Buddhism with different choices and tasks. The heat and flora of South Asia are oppressive but also productive. Food and building materials are plentiful. The temperatures and terrain of East Asian mountains can be harsh. The southern monks must contend with insects and wild beasts. The mountain monks endure the cold. Yet what they share is a domain peripheral to the centers of power and civilization. Their chosen locations speak of a retreat to something more primal. In their view, they are a return to the essential teaching of the Awakened One, the Buddha. In eschewing the ornate paraphernalia of mainstream Buddhism, they seek to embody the pristine. To understand the shared temperament of Ch'an/Zen and the Thai Forest Movement, we must begin with their primary project of awakening which positions them as revivalist Buddhist movements.

Two Revivalist Buddhist Movements

The Thai Forest Tradition and Ch'an/Zen present themselves as reactions against Buddhist formalism and scholasticism. Along with generating innovative forms, they position themselves, at least initially, as revivalist sects, returning to or continuing the Buddha's core teaching through their rigorous practices. Their revolt against mainstream Buddhism shapes

A.R. Lopez, *Buddhist Revivalist Movements*,
DOI 10.1057/978-1-137-54086-7_1

their narratives and identity. The Thai Forest Tradition should be understood as a specific movement within a broader tendency, or general social movement, that has flowed through the Buddhist world since its earliest period. This general social movement, the forest tradition, has stood in contrast and some tension with the scholar monk establishment. The polarity has been at play in Buddhist lands other than Thailand, most notably Sri Lanka. Even during the lifetime of the Buddha, a division appeared between those monks who found a comfortable home in urban centers and their often royally funded parks and those who gravitated toward the rigors of life in the forest. By the time Buddhism was fully established in Sri Lanka, the polarity of scholar and forest monk was set.

The affinities that Ch'an/Zen and the Thai Forest Movement display are not explained through the categories of denomination, culture, or historical epoch. Rather it is a shared stance toward mainstream Buddhism, their revival of enlightenment as a living experience, the demand for self-cultivation, and the role of charismatic masters that form a social crucible cooking up surprising similar fare. The teacher student relationship is the dynamic center line that gives definition to both movements, and it is to this interplay that we turn in the following chapter.

Why a social movement? The study of religious phenomena as social movements has a significant literature. Along with the study of cults and sects, observations of religion as process especially involving the routinization of social actions from an inchoate group to an established church invite a social movement perspective. Yet for intelligible reasons, social movements both in the mind of the scholar and the general public are associated with civil unrest, mass action, and political conflict. Even when the subject matter turns to religious sects and cults or new religious movements (NRMs), the other worldly, individualistic, and mystical groups have been the least studied. This tendency is understandable given the absence of mass constituencies, political programs, formal organization, and, at times, official documents. However, once we broaden the definition of social movement to include networks of relatively few individuals with a shared narrative or meaning system, we are able to approach monks of the forest and mountains as participants in social movements. Here are collectives whose goals and actions are the transformation of the human condition, Buddhist liberation.

The choice of a social movement perspective is based on more than Ch'an/Zen and the Thai Forest Tradition qualifying as social movements. Social Movement theory encourages an understanding of highly motivated agents generating social forms through action. Activity at the margins of the social order and intensely held narratives supersede formal organization and official doctrine. Ch'an/Zen and the Forest Tradition essentially grow from face-to-face interactions that have no official status but are based on choices by the actors. A social movement view brings into focus the telos, or ultimate goal and meaning of actions, and is therefore more suited to understanding individuals and small groups that are not simply replicating officially sanctioned patterns. Human activity is a seen as a project. This notion of project is critical for unpacking the dynamics of the mountain and forest monks. As we shall show, the shared features of the project of the two sects are what "drives" their independently emerging affinities.

The Thai Forest Movement and Ch'an/Zen understood as networks of action constituted by a share project bring into focus the non-institutional aspects, informality, fluidity of their social forms, as well as the vivid personalities that especially characterized the formative stages of the world of the monks and masters of the forests and mountains. Viewed as projects animated by strongly held goals, we can ask pertinent questions. What motivates the actors? What are the sources of legitimate authority? What is the manner of teaching and practice? What constitutes their social cohesion? What are their evolving social forms and doctrines? And as a comparative study, what are the affinities and differences between the Thai Forest Tradition and Ch'an/Zen and how can we account for them? A social movement perspective frames such questions and provides a paradigm for inquiry.

The mountain and forest sects are not only social movements but also revivalist social movements. In what sense can we label these two phenomena revivalist? In the popular imagination, revivalism conjures images of tents on the plains of the American Midwest packed with local enthusiasts, agitated preachers, and hysterical healings. What could be further from the Thai Forest Movement and Ch'an/Zen? In contrast to the almost rowdy mass event of the American mid-lands, the forest monks offer a silent, solitary image. Ch'an/Zen contrasts with its inwardness and disciplined self-control. Whether in the monastery or their natural habitat, forests and mountains, they could not be further in outer form and tone from popular notions of revivalism.

Academic uses of the term, revivalist, are primarily found in conjunction with fundamentalist reactions to globalization, modernism, and secularism. Movements from the Islamic revival in Malaysia to the cults of the native people in North America have all been examined as efforts to revive a traditional culture facing disruption and domination by an alien power.

Revivalism has a strong association in the literature, with fundamentalist responses to a loss of status and power in the globalizing process (Hannigan 1993, p. 2). So in what manner can we dub the Thai Forest Movement arising in a twentieth century expansionist Siam consolidating its control over a chunk of South East Asia and Ch'an/Zen waxing and waning in tandem with the dynasties and empires of the Far East as revivalist?

Like any phenomena, social movements can be labeled in a variety of ways with each designation delineating without necessarily denying other dimensions. Here the term "revivalist" accentuates how both of these social movements imagined themselves in relationship to the larger Buddhist community and to the historical tradition. Turning to the word itself revivalism or its root to revive is to bring back to life or consciousness or in the case of insentient phenomena to return to existence or prominence anything from an industry to a mode of thought. The key is the word stem 're' which indicates a prior condition being made contemporary.

Both the Thai Forest Movement and Ch'an/Zen understand their project as a continuation/restoration of the essence of Dharma/Dhamma in the face of what they considered to be a decayed and distorted Buddhism. In the context of popular and state supported Buddhism, their reactions seem radical. Therefore, we should not confuse them with revivalist impulses that are primarily concerned with the reinforcement of traditional social forms and norms. Revivalism is not necessarily social conservatism. Calls to practice the original message can be threatening to the status quo. Both mountain and forest monks can be seen as radical revivalists who distinguished themselves from their religious environments, albeit in some different ways, to actualize what they understood to be the original project of Buddhism.

The Ch'an/Zen sect is more complex in its revivalism than their forest brethren. The Thai Forest Movement is a recent twentieth century event whose brief history affords easier generalization and whose proximity to its founding impulse makes clearer its revivalists origins.

Ch'an/Zen on the other hand has a longer history draped over the entire Far East. Relations with the state and with the larger Buddhist sangha have varied. Both scholars and practitioner have labeled Ch'an/Zen as stagnant and degenerate in certain phases of its history including its contemporary condition. Nevertheless, the Ch'an/Zen School evidences an enduring self-presentation and *raison d'etre* as a revivalist sect. It is notable that the outstanding teachers of Japanese Zen all positioned themselves as reformers. Dogen, Bankie, and Hakuin all found the institution of Zen wanting and harkened back to the great ancestors of India and China as their true ancestors. Zen's self-definition, despite its organizational collusion with Imperial Buddhism, remained that of a revivalist project bringing forward the essential practice/awakening of the Buddha.

The Thai Forest Tradition has battled accusations that it is a Trojan horse insinuating royal authority in to the restive regions of Lanna and Isan. Certainly after an initial phase of mutual suspicion, the forest monks and the Thammayut sect realized the mutual advantage of a royally supported mission. Forest masters were declared national saints. Major monastic centers were built in the North and North East. Nevertheless, an undercurrent of revivalist tension with the centers of power persists. Informants commented frequently on the debased condition of the sangha. One monk spoke of his distress over the state of the sangha as constituting a barrier to his practices. Others spoke of keeping a low profile with regard to the sangha hierarchy even as they spoke glowingly of teachers and centers that kept their distance from the corruption and maintained a rigorous, pristine practice in remote locales. With pride they noted either the hidden location of their collective boxes or their non-existence in contrast to the aggressive solicitation of funds at other temples. The forest monks continue to hold their revivalist, core values as counter-point to the mainstream, institutional Buddhism.

In labeling forest and mountain monks as revivalist, it is their soteriological project that is identified and chosen as their defining feature. As we shall see, it is this project that gives direction and shape to both movements. Their roughly shared goal and method of pursuit are the source of their remarkably similar features. An understanding of the praxis of these two movements requires recognition of their soteriological project not merely as a static classification or type but as a guiding and germinating dynamic. What drives the masters and the students in the forest and mountains is the thirst for the Buddha's one-taste-Dhamma: freedom. To understand the logic of the Thai Forest Movement and Ch'an/Zen is to acknowledge the paramount place of the soteriological quest, which, in their view, sustains and revives Buddhism's original purpose.

Preview

Our study commences with a historical review of the Thai Forest Movement and Ch'an/Zen (Chap. 2). As a general trend within the sangha, the Forest Tradition dates to the time of the Buddha. The role of the *aranyawaasii*, the forest dwellers, will be traced from the jungles of Northeast India to Sri Lanka and then on to Siam and Southeast Asia. Ch'an/Zen will be discussed in its earliest coalescence in China. In reviewing their genesis stories, their self-view, especially with regards the larger Buddhist community and their primary orientation, will be determined. Both movements are shown to be revivalists with a soteriological-pragmatic project to be carried out through self-cultivation under the direction of charismatic teachers. Next, we examine the teaching tactics employed by the charismatic masters (Chap. 3). These teacher/disciple exchanges are the "living dharma" which constitute the key relationship of the Thai Forest Movement and Ch'an/Zen. The varieties of tactics employed are identified along with their functional consequences for the soteriological project.

The investigation broadens to review the practices and overall style of the two schools (Chap. 4). Core practices such as meditation, alms rounds, work, meals and diet, and the organization of temple life are compared. In style, Ch'an/Zen and the *Kammatthana* (literally a place of practice, but colloquially designating the Thai Forest Tradition) are found to contrast along an individualist/collectivist dimension. What they share is a martial spirit and a framing of the spiritual path as a military campaign. The telos of both movements is awakening, enlightenment. In Chap. 5, accounts of awakening events are studied from a phenomenological perspective, that is for their descriptive content, and structurally, or changes reported in the forms and functioning of the mind. Finally, the role of awakening for the social movement is explored.

The theme of the charismatic teacher is approached in Chap. 6 from the records of the lives of the awakened, both auto and biographical. Universal patterns that replicate myths and rituals of passage and quest are identified along with variations within the genre. Ch'an/Zen and the Thai Forest Tradition are found to develop different templates. Sub-lineages of the Forest School also evidence shifts away from more classical hagiography to more modern, naturalistic renderings. Nevertheless, the value of the teacher's story as critical to the imagined history of the school is affirmed.

Questions of doctrine are examined in the next two chapters. Chapter 7 focuses on the teaching of pure mind. The qualities of pure mind and its related designations as radiant mind, original mind, and natural mind are identified. In light of the controversy within the Theravada, textual sources and contemporary explanations and defenses are presented. In pure mind, mountain and forest monks find common ground. Chapter 8 examines different views of the "view" from the suttas to the recent teaching of the Thai Tradition and Ch'an/Zen. Both schools are found to hold an attachment/pragmatic position on right view underscoring the importance of the soteriological project over fixed dogma. Shifting interpretations of key Buddhist concepts by some forest teachers were also found to imply a reconfigured path and goal.

In Chaps. 9 and 10, our study comes full circle placing both movements back in their historical context. Chapter 9 explores the mix of Buddhist modernism and revivalism. The concept of modernism is clarified and applied. In this context, the controversy over female ordination is discussed. In the concluding chapter, findings are reviewed, and questions are raised about the paradigm of Buddhist studies. A final note offers some reflections on the Thai Forest Movement and Ch'an/Zen from the point of view of a practitioner gazing down two paths that repeatedly intersect and part ways yet seem to head in the same direction.

The Symbolic and the Real

A question that lurks throughout this study is what is the perspective taken on the phenomenon under study. How are we approaching Ch'an/Zen? What is meant by Ch'an/Zen and the Thai Forest Movement? More specifically are they an objective, historical social fact defined by demography, historical personages and verifiable behaviors, and institutional relations with other structures? In short, are we viewing Ch'an/Zen and the Forest Movements as "objects" in a social world much the way an outside observer of a plant or a building might analyze those "objects" or are we grasping them with regard to a different dimension? That alternative view is concerned with how the actors present/imagine themselves. The question and the choice between these two views is a classic concern of the human sciences.

The recognition that human beings act, at least ostensibly, on the basis of subjective meanings, both at the level of interaction and under the collective guidance of meaning systems, culture, and ideology, has been central to a view of social phenomena we can call "symbolic." From this view, it is essential to understand meaning both in how it is articulated and how it shapes the behavior of actors. How do actors imagine themselves and invite others to imagine them. This tradition is a long and varied one reaching back, according to some, to Max Weber but moving forward through G.H. Meade (1967) and symbolic interactionism, Irving Goffman's (1959) dramaturgical model, and numerous other formulations operating from the same premise: meanings are essential to action and are their own realities, sui generis. How actors, individual and collective, communicate themselves is critical. The relevance to social movements is obvious. Social movements call upon us to do more than identify their demographic and institutional structures. They ask that we enter the world of symbolic meanings that motivate the actors.

This study neither requires nor will attempt to resolve or advance the discussion on the relative value of each view. Both perspectives are involved. At times we will look, for example, at what "really" was the state of Zen as an objective institution, but at other times, the concern will be with how Zen presents itself. What is critical is which point of view predominates at different moments. Both the symbolic and the object perspectives are relevant to the inquiry, but it is the former, the symbolic, which is most often pertinent to this inquiry. In the final analysis, the focus is how Ch'an/Zen and the Thai Forest Movement present and understand themselves which in turn significantly shapes their actions.

Exploring the micro-interactions of teaching tactics, the phenomenology of awakening experiences and the constructed auto/biographies of masters gives center stage to the world of meanings. Yet the "real" story of Ch'an/Zen and the Thai Forest Movement (how they behaved behind the imagined story, and their inter/intra-institutional relations) remains relevant, if only because it contrasts with the constructed nature of their imaginings and puts into relief narratives that seek, at times, to project an alternate story. Finally, the emphasis on the symbolic is chosen as it is in this dimension that the two movements articulate commonalities and commit to meanings that generate their inner dynamics and public presentations.

Research

Data is derived from multiple sources which triangulate the subject. There are primary materials, printed, audio, and increasingly videos produced by the movements and its luminaries. Secondary sources include studies of Thai Buddhism, Far Eastern Buddhism, and eyewitness and personal accounts of the forest school and its mountain-based cousins. Solid anthropological and historical studies are supplemented by autobiographical and biographical materials. Ch'an/Zen and the Thai Forest Tradition, given the intensely personal nature of their involvement, have inspired a number memoires and works of personal reminisces.

Finally, there is the author's field research consisting of participant observation and open-ended interviews. Unlike other fieldwork research projects that commence with a specific study and close with its conclusion, the research that supports this study began long before the project was conceptualized. Data gathering began 40 years ago when the author attended his first *sesshin*, Zen meditation retreat. Contact with the Thai Forest Movement began some 20 years ago. Primarily, this involvement consisted of participation in retreats or solitary practice at forest-affiliated centers. Addition contact came through educational work in Thailand that introduced me to such Forest Tradition luminaries as Ajahn Sumedho, Luangpor Viriyang, and Ajahn Plein, as well as numerous third-generation teachers.

Open-ended interviews were carried out with monks from both sects, including Gouyuan director of practice at Dharma Drum Mountain in Taiwan, Amaro Bhikkhu, Pra Panlop, and others of the Forest Tradition. However, certainly of at least equal value were the informal exchanges while sitting together in a temple kitchen, comments passed during a work break, or on a forest stroll. Often these moments surpassed my set piece questions in opening the researcher to new perspectives.

Technical Issues

Most texts present technical difficulty, and this work is no exception. Spelling and capitalization present problematic choices often devoid of satisfying alternatives. A work involving Thai Buddhism is especially problematic as it confronts the writer with the overlay of two notoriously awkward lexicons: Buddhism and Thai. Further compounding the issue is managing of two branches of Buddhism with different root languages, Pal and Sanskrit. Finally, with the migration of Buddhism across East Asia, there is opportunity for name changes and disputed pronunciations.

Examples of the above that required resolution are the Buddhist term "dharma" in the Mahayana and "dhamma" in the Theravada tradition. The problem is double. When to use the Sanskrit and when to use the Pali version is the initial issue. Next is the question to capitalize or not to capitalize. Fortunately, this problem has confronted scholars for over one hundred years, and their solution is adopted here. As an adjective (dharma talk), small case is used. When referring to the teachings or ultimate reality, the word is capitalized. The term sangha demands similar choice. When working as an adjective (e.g. dhamma talk), it is not capitalized while it receives capitalization as one of the Three Jewels of Buddhism: Buddha, Dharma, Sangha. As for dhamma vs. dharma, I have tried to use the form that is spoken in the tradition being referenced. Ch'an has dharma and the Kammatthana has dhamma. Sometime both are used (dhamma/dharma) when the reference is inclusive of both schools. This logic is also at work in the use of Ch'an or Zen. There has been recently the assertion that Zen was the original pronunciation while Ch'an was a soften term introduced by the court. This is not certain. What is certain is that Chinese from Shanghai to Chinatown, NY, use the term Ch'an to identify their tradition. So be it. Again to indicate inclusivity I have used Ch'an/Zen. When only referencing the Chinese or the Japanese tradition, I use either Ch'an or Zen.

Lastly, there is the complexity of Romanized Thai and Thai pronunciation of Pali terms. The Thai language has resisted a standardized transcription into roman letters. Names, place, and actions all seem to find idiosyncratic forms. Even as important a name as Ajahn Mun/Man offers choice. On the basis of no particular wisdom, I have chosen to conform to the most frequent usage. The use of either Pali or Thai has followed the logic of dhamma/dharma. When discussing the phenomenon in as Theravada, the choice is Pali. When the context is the Thai Tradition, the Thai term seemed appropriate. Frequently both are used for clarity.

In text, distinctions are often not easy to maintain. There is perhaps a single default principle. Despite the warnings of the Buddha and his children that all is inconstant, I have chosen to worship at the shrine of the god of consistency. Hopefully, this will provide coherent passage for the reader.

Summary

Ch'an/Zen and the Thai Forest Tradition were examined from a broad and introductory overview. They were both categorized as revivalist Buddhist movements. Revivalism was defined generically as any project at restoring or maintaining a way of life or system of action that is justified by the original or essential intentions and practices of their traditions. It was noted that this does not necessarily entail social conservatism. Rather in the two cases under study, it may be a radical repudiation of the predominant social practices in the name of a purer and more authentic truth.

In contrast to the metaphor of their distinct habitats, mountains and forest, an affinity between these two movements was posited. In addition to their chosen marginal terrains, a wilderness beyond the civilized center, they share commitment to Buddhism as a quest for a proximate awakening through self-cultivation. This shared stance will be shown to germinate multiple identical features in two Buddhist schools that are otherwise separated by formal doctrine, geography, and historical epochs.

A brief preview was offered of forthcoming chapters that will unfold and specify the points of similarity and difference between Ch'an/Zen and the Thai Forest Movement. Finally, the approach to the subject at both the conceptual and the empirical levels was noted. Conceptually, it is the symbolic or meaning dimension especially the self-presentations and self-imaginings of the movements that will be accented over and above a perspective of historical or cultural realism. While the latter will be recognized, it is their working narratives that will be seen as shaping their practices and even their evolving doctrines. The data for this study is drawn from multiple sources that triangulate the material under investigation. Both movements have extensively documented their narratives. Along with written and recorded materials, contemporary materials include extensive audio/video data. Secondary sources are also abundant but not of a comparative nature. Field research, it was observed, took two basic forms. Participant observations research including retreat participation and living at practice centers over a period of several decades and in-depth interviews with teachers and students in both traditions.

References

Hannigan, J. (1993). New social movements theory and the sociology of religion. In W. H. J. Swatos (Ed.), *A future for religion* (pp. 1–18). Newbury Park: Sage.

CHAPTER 2

Streams and Sources, Forests and Mountains

Introduction

The Ch'an/Zen movement and the Thai forest school are separated by almost 1500 years in time of origin, by 1000 kilometers in location, different cultures, languages, and of course by religious denomination. Study of the Thai forest movement and its relationship with Ch'an/Zen both in its classical T'ang dynasty origins and in its contemporary manifestations in China, Japan, and the West is a comparison of Buddhist schools that have developed in almost complete isolation and ignorance of one another. Yet they warrant comparison. Only in recent years has the curtain of separation parted permitting a trickle of translations to pass through. In the twentieth century, Western students (and some Asian) have become conversant with both movements, through direct contact in retreats and conferences. The absence of communication during their formative periods makes their affinities all the more fascinating and revealing of critical dynamics. Our investigation begins with an overview of their socio-political settings, their genesis stories and self-narratives, and their orientation as revivalist movements challenging the Buddhist world around them.

The Genesis Stories

The Ch'an story begins with the legendary Bodhidharma and his interview with the Emperor Wu, the embodiment of state supported Imperial Way Buddhism. After disembarking on the coast of China from his South

A.R. Lopez, *Buddhist Revivalist Movements*,
DOI 10.1057/978-1-137-54086-7_2

Indian homeland, he makes his way inland to the Emperor's court. There in an iconic interview with the monarch he defines the Ch'an/Zen movement's distinctive stance toward doctrine and practice. Asked about the quantity of merit acquired by the Emperor through his considerable donations to institutional Buddhism, Bodhidharma answers, "Nothing." What is Dharma/Truth the Emperor demands to know. "Vast space, nothing holy," the foreign monk replies. With responses that are as puzzling as they are impertinent the Emperor demands to know who stands before him. "Don't know," says the first Ancestor of Ch'an. Turning on his heels he marches off to the environs of Shaolin temple where, as the story says, he sits for nine years awaiting his first disciple. Significantly, he sits outside the temple. He is related to Buddhism but not inside the ecclesiastical hierarchy. He meditates rather than translates. He faces not a Buddha statue but the wall, a wall gazing practice, *puan kuan* (Ch.) that embodies the radical directness emblematic of his sect. Whatever its historical veracity, we have the genesis tale of a new movement whose pith slogan was to be "A special transmission outside the scripture, a direct pointing at Mind." With this story, a new Buddhist school, Ch'an/Zen, is branded (Ferguson 2012, pp. 15–28).

The Thai Forest Movement's declaration is less confrontational but equally dismissive of royal authority. Ajahn Mun having been appointed to a post at Wat Chedhi Luang, a temple of the royally founded Thammayut sect in Chiang Mai, disappeared without formal notice. Leaving behind a brief note and his regalia, Mun walked into the forests north of Chiang Mai where he practiced in the famed forests and caves of Chiang Dao. In this wilderness Mun, according to his disciples, attained ultimate Release, Nibbana. When his student, Ajahn Thate, sought out his master at Chedhi Luang, he was met with the hostility of temple officials who referred to Mun "with dismissive contempt" (Tiyavanich 1997, p. 265). To a Thai Theravada Buddhist world that had just recently undergone state-imposed routinization, and with the widespread belief that Nibbana was unavailable in the modern age, Ajahn Mun's attainment was revolutionary. Like his brother Bodhidharma, Mun snubs mainstream authority abandoning the temple for the cave. The Forest Tradition, the Kammatthana, was now connected to the original root, the Buddha beneath the Bodhi Tree. With Ajahn Mun's claim that Nibbana was available and not in the wats, the court, or the colleges, but in the forest, the Thai Forest Movement as a distinct twentieth century phenomena is born.

Forest Monks from Bodhgaya to Colombo

The forest dwelling monks, *aranayawaasii*, date to the earliest days of primitive Buddhism. One could say that Buddhism came out of the forest and has had a continuing penchant to return. Thai Buddhists are frequently heard to claim that the Buddha was born in the forest, was enlightened in the forest, and died in the forest as irrefutable evidence of the inexorable connection between dhamma and forest. The forest is both the historical and symbolic matrix of Buddhism. Even as monks in India began to live in large, stone monastic complexes, they carved into the walls and contoured the architecture with forest motifs that evoked their prior jungle habitats (Schumann 2004, p. 175).

The forest is more than a romanticized landscape. Forest monks, as a distinct trend within the sangha and a counter-point to the scholar monks (*pariyat*), date to the Buddha's lifetime. While there were sangha members who preferred residences near the nexuses of power and wealth, others wandered in the forest as did the Buddha early in his career. Although the Buddha is often recorded as interfacing with kings and merchants, we also read of his encountering his monks deep in the forest and extolling their way of life. On several occasions during the course of his 40-year ministry, the Buddha retreated into the woods for a lengthy solo retreat (Schumann 2004, pp. 195–196). Buddhism has been analyzed as an urban religion of North India's mercantile society, nevertheless it retained a lure for the forest and a sufficient number of ordained persons followed that call to constitute a forest tradition.

This polar tension that was to mark the sangha seems to have been present since the inception of the sangha itself. These tendencies have been variously labeled: forest vs. city monks, meditating monks vs. scholar monks, wandering vs. sedentary monks, or forest dwelling vs. village dwelling. Although the designations do not perfectly align and have descriptive variation, they, nonetheless, refer to roughly the same polar currents that flowed through the sangha from northern India, to Sri Lanka, to Southeast Asia, from Bodhgaya to Bangkok. The *Auguttara Sutta* contains passages that have been interpreted as the Buddha's effort to bridge the gap between the meditating forest dwellers and the more intellectual monks of the city with the Master finding value in both. He speaks of "dhamma experts" and meditation monks thereby recognizing the two camps. The gap persisted between the two, but forest monks were explicitly acknowledged faction. By the time Buddhism had reached Sri Lanka, the forest monks were an official component of the community with their own centers, high clerics, and recognized religious specialty (Carrithers 1983, p. 168).

Buddhism's Janus face is evident in Medieval Ceylon's double-sided sangha. Although formally a single community, the institutions of the forest wing were recognized. Here it must be noted that they had their own monasteries or hermitages and were not the hermetic wanderers of popular imagination. What characterized their communities was asceticism and meditation practice. When matters of purity and discipline became monarchial concerns, it was to the forest section of the sangha that the king would turn. After recovering from the disaster of the Cola Invasion, King Parakkamabahu the Great called on the head monk from a large forest/meditation hermitage to interpret the *Vinaya*, the Disciplinary Code. The *Mahavamsa* of the sixth century records royal gifts to the rag-wearers and forest dwellers. A similar acknowledgment of a de facto dual sangha seems to be expressed in Burmese designation of those who "walked alone" and those who "walked with the many" (Carrithers 1983, p. 171). Collectively, the forest monks of Sri Lanka, Burma, and Thailand constitute a general social movement active in the main centers and throughout the nearly 2500 year history of Theravada Buddhism. Within this larger orientation, there developed specific social movements, that is both lineages (*nikayas*) and sometimes formal organizations giving a particular articulation to a broader language. Recent centuries have witnessed such developments in Sri Lanka and of course in Thailand.

The story of Sri Lankan Buddhism is one of revival and resistance to both Hindu and Western penetration. Thai monks were imported at one point to provide the requisite numbers for ordination for the devastated sangha. Recent centuries featured concerted campaigns to re-establish the forest tradition. In the twentieth century, a fertile partnership of two monks, Jinavamsa and Nanarama, succeeded in planting a new forest hermitage movement. Unlike Thailand, this was an intentional effort. Jinavamsa wrote a series of newspaper articles which attracted lay support. His alliance with Nanarama was the critical jumping off point. With land, finances, and a plan, all that was needed were monks. The solution: advertise in the newspapers! To Thailand's Ajahn Mun, this would have been fantastic, perhaps humorous, certainly redundant. Yet it worked and Sri Lanka today has an extensive network of forest hermitages (Carrithers 1983, pp. 191–193).

THE THAI FOREST MOVEMENT: FROM SUKHOTHAI TO BANGKOK

The Thai movement is an ongoing tradition pre-dating the Siamese state rather than an intentional startup in the modern era. Instead of a plan, it was a transplant from Sri Lanka. Although our evidence is fragmentary, it seems likely that forest monks were a part of the landscape since the arrival of Singhalese missions. Earliest evidence from the Sukhothai period suggests a sangha formation much like Sri Lanka's. There were two sections: a right hand of *khamawassi* or town dwellers and a left hand of *aranyawaasii* or forest dwellers. A local informant showed the researcher stone structures which he claimed to be the meditation cells of forest monks set in the wooded hills to the west of Sukhothai, while the main eastern gate close to the palace was the residence of the court dwelling monks. In dire times when the services of the *khamawaasii* faltered, the king's entourage would walk west to confer with the forest monks and their chief, entitled the Wanarat or Paa Kaew. In short, the forest monks were a defined and recognized presence (Taylor 1993, p. 27).

The prominence and status of early period forest monks was progressively eroded during the following Ayutthaya and Bangkok Periods. By the reign of Rama I, the forest monks were no longer an official division within the sangha. While Rama III, Mongkut, was establishing a new *nikaya,* Thammayut, to inject rectitude and scholarship into the sangha, his Burmese contemporary was, in contrast, urging monks to take to the forests. With the reign of Chulalongkorn, passing exams in Pali on the way to ecclesiastical summits had replaced ascending through the meditative jhanas on the way to Nibbana. The forest movement had been de-institutionalized. Yet forest monks survived and even showed a presence in Bangkok. Despite the march toward the bureaucratization and the subordination of the sangha to the monarchy' s will, there was an ambivalent dance between the Bangkok powers and the forest dwellers. Individual masters would impress a king with their attainments, and forest monks would camp on the outskirts of Bangkok at times attracting aristocratic patronage.

The relationship of the Bangkok elite to the forest monks is ambiguous and evolved over time. The specific forest movement traced from Ajahn Mun crystallized out of a milieu that existed for centuries but may have been further paradoxically energized by Bangkok in its suspicion and rejection on the one hand and by its support of virtuoso monks on the other. King Mongkut, Rama III, after a lengthy career in the monkhood disrobed and as Siam's new monarch toured the temples of his kingdom and was aghast at the conditions he encountered. As an intellect with rationalist leanings and a monarch with an appetite for central control, he was disappointed on two counts. Confronted with village monks who knew more magic than dhamma and were entwined in a parochial world, he determined to correct both conditions by establishing a new order. His Thammayut group would live by the Vinaya and know the Dhamma of the Pali Suttas. The Thammayut were to be the royal shock troops in the Siamese Sangha much as the Jesuits were for the Vatican in its struggle with heretics. However, the new group did not ascend to dominance until the reign of Rama IV, Chulalanlorkorn. With his monk half-brother at the helm, the sangha was restructured. The Sangha Reform Act 1902 mandated a series of extensive changes. Pali exams were standardized, a catechism was written, the sangha was bureaucratized and centralized, and regional resistance was smothered (Watt 1984, p. 216). The appeal of the Thammayut to Mun and his associates is perhaps based less on the monarchy' s designs than on an attraction to a revived dhamma of liberation and the ascetic life style. Most of the monks who followed Mun were either Thammayut or converts to Thammayut, but this was not universal. Some disciples were Mahanikai and chose to remain with their original affiliation. Ajahn Chah, undoubtedly one of the most influential forest masters, remained a Mahanikai monk.

Furthermore, a closer look at the Thammayut Forest monks revealed many "non-Thammayut" traits. Sao and Mun were deeply imbedded in the Lao tradition. Others found commonality with Cambodian customs of their native villages. From chanting styles to meditation practices to dress, the first generation of Mun's forest monks may have been closer to the village monks opposed by the Thammayut hierarchy than the educated monks of Bangkok.

What is clear is that the twentieth century Thai movement of Ajahn Mun did not arise ex nihilo. (In these pages, the term Thai Forest Movement will refer to the specific formation that traces its lineage to Ajahn Mun.) In addition to the anonymous forest teachers scattered in remote areas across Thailand, Myanmar, and Lao who Mun sought out for guidance, two historical figures influenced the course of Mun's career: his friend from his home region, Phra Ubaalii, and his elder brother monk, Ajahn Sao.

The standard narrative about Mun and the Forest Movement is that it represented a blending of Thammayut discipline and forest praxis. The situation, however, is more complex. Mun's preceptor and lifelong shield against ecclesiastical suspicions, Ubaalii, was a Bangkok-based monk with an eccentric personality that might be better thought of as found in the wilds. Ubaalii, known for his uncompromising direct teaching, was reported to present himself in rags despite his high position and to carry his monk's bag inside out to show that it is what is inside that matters. When placed under "temple arrest," he hung a bag of yams outside his window playing on the Thai word for determination (Taylor 1993, p. 57). Apparently, Mun's affinity for eccentric monks and their attraction to him pre-dated his years of wanderings. Nor were all the robes of Thammayut monks as tightly wrapped as the official version would have us believe. Ajahn Sao, on the other hand, was a true forest comrade. Senior to Ajahn Mun, he had already established his reputation in the northeast as a meditation master. Sao's style was, if possible, even more solitary and reserved than Mun's. Together, they lived the austerities of the wandering ascetic or *dhutanga* or *thudong* (Th.) monk. Sao left no written teachings, but is recognized as a forest master in the tradition associated with Mun.

An additional figure, Maha Juum, should also be noted. He was a ranking scholar monk who lent credibility to Mun and his practice centered disciples. Like Ubaalii, Juum was an intermediary between the Thammayut hierarchy and the forest monks. While Sao and Mun were the forest masters, it was Ubaalii and Juum who offered translation to and protection from a world the two dhamma brothers had distanced. Taken together, these four personalities account for the formation and the positioning of the Thai Forest Movement in early twentieth century Siam.

Mun's career has been parsed into four phases. From 1892 to 1915, he wandered mostly in the northeast culminating in his attainment of non-returner and the commencement of his teaching career. A second phase of wandering 1916–1928 even took him to Bangkok. The third phase of wandering in the north was marked by the attainment of the path/fruit of arahantship. The concluding period was nine more years of wandering in his home northeast (Taylor 1993, pp. 104–105). Born in the northeast in 1870, Mun was according to Tambiah (1976, p. 14), a regional representative of a universal ideal, that of the arahant. He was understood as a contemporary display of the qualities to be found in the classic arahants of the Buddha's time: Sariputta, Mahakassapa, and Upali. His environment was the thinly populated but densely forested northeast, Isan. The area was remote, and with the absence of telegraph and railroad, relying on village " runners" communication, let alone transport, could take weeks. Physical distance was compounded by cultural distance. The area spoke a Lao dialect and ruled by local elite not connected to Bangkok. The inaccessibility that to the crown must have been seen as an obstacle to its nation building was a blessing to Mun and his wilderness abiding comrades.

Mun when not traveling alone moved with small groups of disciples stayed at habitable caves or in forest *samnaks*, hamlets that hardly rose to the stature of monasteries. Mun's constant movement not only suited his style, but allowed him to evade appointments to administrative posts from the sangha hierarchy. Although maneuvered into abbotship at Wat Chedhi Luang in Chiang Mai, the most important Thammayut temple in the north, he sought escape by every means. As mentioned, he fled leaving behind his insignia with a note indicating here was Phra Kruu Mun, his official title. Mun counseled his disciples to do likewise and effusively praised solitude and nature (Mah Bua 1995a, p. 224). Mun's escape to the forest echoes Bodhidharma's trek to Shaolin and its nearby cave.

The Thai Forest movement has a strong regional identity. The forest movement's regionalism is in apparent contradiction and actual tension with its ties to the Bangkok center. Taylor (1993, p. 41) lists by region of origin forest monks of the first two generations. The result shows an overwhelming preponderance of northeastern or Isan monks. Out of the 64 monks included, 53 hail from this geographically large but under populated and undeveloped area (Taylor 1993, p. 109). While this might make them ideal agents for the center's imperial designs, there is no evidence to show that they were chosen and trained by any ministry of interior or defense. Rather these monks are an expression of their local context, a hinterland on the Thai/Lao frontier. Lao had been long recognized sometimes enthusiastically and sometimes with suspicion as a Buddhist land of the supernatural and of meditation prowess. Even a famous monk and court favorite like Somdet To was reputed to spend considerable time in the baan Lao, the Lao neighborhoods, of Bangkok (McDaniel 2011, p. 39). The people of Isan are largely held in contempt by the Bangkok elite and their aping middle class. They are consigned to clownish, demeaning roles in national media much like the treatment afforded African Americans and Latinos in past decades in the USA. Yet individual monks from this region, mostly affiliated with the royally founded and funded Thammayut sect, would become the crown's icons.

The Forest Movement fought a war of resistance on two fronts. While it has been argued that Mun's enterprise was eventually co-opted into the state's nation building project with its luminaries inducted into the pantheon of national heroes, the early history of the movement indicates a revivalism in tension with dominant brands of dhamma and its institutions (Taylor 1993, pp. 40–41). The skepticism of the official Sangha was more than matched by the hostility of village monks and at times villagers. The forest monks were a third camp distinct from metaphysical-state Buddhists and the popular-village Buddhism. If the wandering that took monks beyond the realm of ecclesiastical supervision raised concern, then the forest monk's repudiation of the non-soteriological popular Buddhism of spirit worship and merit gathering evoked animosity and even violence from locals. Effort was expended to tame the spirits and pacify the villagers, as well. The ritual paraphernalia and altars of spirit worship were destroyed. As a movement seeking a return to or extension of the ways of the early sangha and its enthusiasm for liberation, the Thai Forest Movement found itself at odds with both bureaucratic and popular Buddhism.

Individual motives and institutional functions need to be separated. In Isan, Northeast Thailand, a culture of intense religious motivation joined with a vast forest wilderness to produce a generation of spiritual supermen. That this movement was effectively incorporated into the state's nation building project complicates the analyses, but does not negate the indigenous and subjective factors that fed a movement of Buddhist revival. If it were not for the membership of forest masters in the Thammayut sect, the Forest Movement's underlying similarity with other revivalist reactions worldwide would be more apparent. Revivalism arises at the margins from Pakistan's northwest frontier to Mexico's Yucatan. The Thai Forest movement shares this marginal origin but with a unique mix of center affiliation and inspiration from a world religion.

Bodhidharma and His Children

The genesis story of Ch'an/Zen introduces all the main themes of its revivalist stance and the style of communication and practice it was to elaborate in the coming centuries. The scene no doubt edited for its signature effect declares a teaching that refutes establishment Buddhism and displays the founding Master's personal fearlessness in the face of the Emperor. A Buddhism of merit making, a dharma that can be conceptually grasped, and even a personal identity are denied. Bodhidharma's departure from the court and his subsequent choice of residence sketch a key theme of early Chinese Zen: eschewing of any proximity with Imperial power. Chronicles of Bodhidharma's career report that he avoided locations associated with state power preferring to move about from monastery to monastery and more significantly to reside beyond their walls. While there is debate as to which cave was the sight of his famous nine year practice of wall gazing, his choice of a natural mountain site over a monastery is symbolic of early Ch'an spirit of vigorous independence. Daoxuan, an early chronicler, tells us that while "those who came to study and honor Bodhidharma were like a city," the blued-eyed barbarian "would not remain in places of imperial sway" (Ferguson 2012, p. 138). Evading Imperial control by refusing court invitations and seeking out remote locations is a family trait that Bodhidharma passed on to his offspring. Hiuke the second Zen Ancestor, as well as Sengstan and Daoxin, the third and fourth ancestors respectively, all refused the Empower. Evidently at patience's end, the son of Heaven demanded Daoxin's presence four times with the last request accompanied by the demand for his head should he refuse. Daoxin bared his neck. The Empower relented with a grudging admiration (Ferguson 2012, p. 142). Even the last of the Ancestors, the famous Sixth, Huineng, upon being presented with an Imperial request for his presence forwarded by the head of the Northern Ch'an School, Shenxiu, followed his ancestor's recalcitrant ways

Ajahn Mun and his followers were aranyawaasii, forest dwellers, Ch'an teachers were mountain men. As the forest was not only the physical environment for practice but assumed a symbolic significance, so the mountains in the great curve of the Yangtze River offered not only distance from the court but were fused with the identity of teacher and the teachings. Teachers assumed the name of their mountain abode, leading on occasion to humorous confusion between the man and the mountain. Perhaps no instance is a more famous example of the oneness of person and place than the name of Zen's foremost poet, Cold Mountain, Han Shan. Daoxin the Fourth Ancestor's dharma was known as the Eastern Mountain Teachings. Man, mountain, and dharma were a single entity; remote, imperturbable, and towering over the conditioned human world.

Here in the cloud shrouded peaks so loved in Chinese landscapes, Ch'an/Zen continued the native Chinese tradition of hermits. Some evidence has emerged suggesting the T'ang Dynasty monks lived in mountainside huts. The earliest generations of Ch'an were peripatetic. It is believed that only with the fifth Ancestor, Hongren, was a considerable, stable monastery created. With Hongren's successor, the Ancestor of the Northern School, Ch'an centers were established in the two capitals, thereby shifting Ch'an's center gravity out of the mountains. Two contemporary studies by Westerners confirm the continued existence of the hermit tradition. Bill Porter's *Road to Heaven* (1993) puts into print the stories of today's hermits, while Ted Burger's work, *Amongst the White Clouds* (2007), puts the lives of the continuing existence of Buddhist hermits on screen.

Even when not as extreme as their hermit brothers, Zennist displays a penchant life beyond the monastery walls. Shitou, a Tang master and the author of several classical Zen texts still chanted in Zen centers, preferred, like Bodhidharma, to be outside the temple. His name, Shitou, Rock Ledge, derived from the natural formation that overhung his grass hut. Hongren, the Fifth Ancestor of Ch'an, extolls the mountain location. Here, the large trees necessary for monastery construction are to be found. However, the preference runs deeper. He advises, "In studying the Dharma one should find refuge for the spirit in remote mountain valleys, escaping far from the troubles of the dusty world. People should nourish their nature in the deep mountains, staying away from the affairs of the world for a long time" (Ferguson 2000, p. 31). It is not accident that Sheng Yen, a contemporary Ch'an master, established Dharma Drum Mountain perched above modern Taipei. To this day, the investiture of a Zen abbot is referred to as the "Mountain Seat Ceremony." Zen passes up no opportunity to advertise itself as the mountain sect even when its location is closer to the tops of Manhattan skyscrapers than the peaks of Wutai Shan or Fuji.

Yet it is not only the evasion of Imperial involvement or the school's remote location that signals Ch'an/Zen's revivalist project. Rather the essence of revivalism is found in the school's very name: the meditation sect. The attitude toward practice/enlightenment, the soteriological project and its pursuit, distinguishes Ch'an from its competitors. Bodhidharma not only proclaims what true dharma is not and what it is and how his lineage realizes it. "Not relying on scripture, directly pointing to the nature of mind, realizing Buddha" is emblazoned on the banner of Ch'an/ Zen. Bodhidharma is insistent that "only by observing the mind" can one achieve liberation. He dismisses scholarship and ritual, the staples of Chinese Buddhism, as inadequate vehicles for liberation. Wall gazing meditation is the highest Mahayana practice he declares. Students are urged to single mindedly look at the nature of mind. In a remarkable series of passages in the *The Breakthrough Sermon* (Red Pine 1987, p. 9) attributed to the beaded founder of Zen, he redefines all the core Buddhist practices and attainments as externalizations of the inner attainments of meditation. Bodhidharma's Buddhism is not outer forms but an inner disposition. This interiorization of the meaning dharma practice recalls the Buddha's similar maneuver with regard to Brahmanical rituals and the forest monks' penchant for subjectifying ritual. We should go to the wat every day, Ajahn Fan advised, but then he added "the wat is in the heart" (Taylor 1993, p. 155). The Heart/Mind is what matters as Phra Ubaalii with his inside out monk bag seems to say.

Along with the emphasis on meditation and the equation of mind with the highest dharma, is the downgrading of the significance of scripture. Huineng, who along with Bodhidharma is the most prominent of the early masters, asks rhetorically, "Why waste effort seeking metaphysical ideas?" Only by observing the mind is wisdom found (Ferguson 2012, loc. 1114). Yet we should recognize that Zen did not reject scripture per se. Rather it was the reliance on scripture over direct experience that drew criticism. The manuscripts from the Tungshan Caves, that tell us much about early Ch'an, are replete with scriptural references. An identical tendency is found in the Forest Tradition. In Ch'an, like the Thai Forest Movement, the use of scripture is supportive and adjunctive to the unmediated knowing of Dharma/Dhamma. Liberation comes through the direct apprehension not mental conceptualization or scholarly endeavor.

Although Ch'an/Zen boasted it was not relying on scripture, it did have its preferred texts which demonstrate its revivalist commitment to practice/attainment. The scriptures can be described as psychological rather than metaphysical. In subsequent centuries, Zen teachers would become interested in the more cosmological treatises of Mahayana Buddhism, such as the *Avatamsaka* and *Lotus Sutras*. However, early Ch'an preferred writings that conceptualized its favorite themes: mind and meditation. The Yogacara or mind-only teachings on the nature of perception and its model of a multi-layered consciousness suited the Ch'an obsession with the direct knowing of mind. The *Lankavatara Sutra* which contains Yogacara teachings was such a document and the first generations of Ch'an were so identified with its teachings that they were called Lankans (Suzuki 1978, p. 25).

The Thai Forest Movement sees itself as the inheritor of a way of life, a practice, and a teaching emanating from the Buddha himself. The Zen tradition claims even more. Each day, Ch'an/Zen monks recite the lineage names of a queue masters extending back to the historical Buddha and beyond. The empirically dubious assertion of this 2500 year one-to-one transmission of the awakened mind testifies to the school's revivalist self-conception. Other sects have strayed, but Zen is the living expression of Mahakasyapa's smile that knew beyond words the truth of the Awakened One. To realize Zen is to entangle eyebrows with the ancestors they say. The Buddha's mind is transmitted in an exclusive line running west to China and now even to East Los Angeles. Taken together, the insistence on direct knowing, the suspicion of pure scholarship, the claims of exclusive inheritance of the Buddha's mind, and the rejection of the Buddhist establishment, all denote early Ch'an/Zen's revivalist politics.

The Central Contradiction of Buddhism

Investigation of the Forest Tradition and Ch'an/Zen and their ambivalent attitude toward establishment Buddhism lays bare Buddhism's fundamental dialectic. The central tension or contradiction within Buddhism is between its world rejecting tendency and its world accommodation. While Weber understood early Buddhism as world rejecting (Weber 1993, p. 269), later Buddhist scholars have noted the Buddha's concern that the Sangha be interdependent with the lay world through, for instance, *pindabat*, or alms round. This point was forcefully conveyed to the researcher in conversation about the Chinese hermit tradition with a forest. Extended isolation was "dangerous" because it weakened the monk/lay connection. At most forest monks could survive three days without food. This assured their ongoing involvement with the lay community. Village monks in Thailand often do planning and directing for collective work projects. And of course, the omnipresent making merit (*tham boon*, Th.) is a very this-worldly exchange of intangible goods, spiritual merit, for tangible goods and services, food, labor, money. In China, ceremonial functions, especially funerals, have been a source of consolation for the laity and revenue for the sangha and also constitute a significant participation in the lay life cycle.

In addition to the everyday functions of Buddhism, macro level social and historical analyses have argued that the Buddhist project provided a workable ethic of reason and constraint for the new urban classes of northern India. Trevor Ling (1973), for instance, understands Buddhism as a total civilization, a potential he believes to be inherent in all world religions. The view of these scholars has come full circle from Weber's interpretation of Buddhism as other-worldly. This new thesis may be stated as Buddhism and the Spirit of Mercantile Capitalism. Nevertheless, despite this corrective interpretation that recognizes the worldly functions of Buddhism, the dual nature of the religion continues to exist and be noted by researchers such as Melford Spiro. In his classic work on Burmese Buddhism, he distinguished "nibbanic" from "kammic" Buddhism, a creed of salvation from a creed of worldly improvement (1970, p. 31). Despite criticism for making too sharp a division, Spiro's categories acknowledge that Buddhism is two sided. Buddhism is both this-worldly and other-worldly.

A Buddhism facing in two directions indicates a fundamental tension within the tradition. An undeniable other-worldliness persists and nowhere more evident than in the Forest Tradition and early Ch'an. What Buddhism aims for and how it survives are two different questions with two different answers. The root event of Buddhism and the apex of its soteriological promise is an inward occurrence, awakening, which is pursued through an often solitary, ascetic discipline. The core purpose (and praxis for some) of Buddhism is other-worldly, yet Buddhism performs complex social functions. Buddhism is not either/or but both/and. Buddhism's story is the containment and balancing of this dialectical tension, one pole of which is embodied in the early phases of Forest Tradition and Ch'an/Zen's.

This central tension in the Buddhist tradition is only recognizable if both poles are acknowledged. To read only the Buddha's *sasasana* or message is to miss the social functions that Buddhism plays from Chinese funerals to Thai alms rounds. Likewise, to see only the cultural context is to miss the transcultural objective in the Buddha's project. Some studies have emphasized "total culture" and the unique gestalts of different "Buddhisms." While a healthy redress of a reliance on text only studies and their transmission to Western audiences, this approach fails to capture the importance of the core enterprise of Buddhism which is so vital to revivalist movements. The study of Buddhism can be over-contextualized at the expense of its attempt at a universal message. Without the other-worldly soteriological, Buddhism's multi-facets are ignored in favor of a one-dimensional normative Buddhism. Buddhism is not only a force for consolation and social cohesion; it is also be a challenge to this world.

Both the Forest Tradition and Zen find support for their revivalism in the Dhamma/Dharma as characterized in the oldest texts. Among the classically noted six qualities of Dhamma are the following three: *Akaliko*, timeless and immediate; *Ehipassiko*, come and see; *Paccattam veditabbo vinnuhi*, to be personally known by the wise. The experiential aspect is undeniable. The intimate vibrancy of Dhamma is clearly not a product of

German romanticism or San Francisco Bay Area counter-culture. Dhamma was not a matter of belief or scholasticism. It was to be directly known by the wise. The quest to know Dhamma, although buried beneath massive monasteries, weighty texts, and obscured by the incense of elaborate ritual, periodically emerges in revivalist movements. These revivals express an authentic although not exclusive version of Buddhism. Bodhidharma's slogan of revolt was "A direct pointing to the nature of Mind." In this century, a leader of the revived Sri Lankan forest movement, Jinavamsa, placed a sign outside his training center that declared, "Buddhism still leads to Nibbana" (Carrithers 1983, p. 222). These pronouncements find support in the core teachings and represent one pole in the Buddhist dialectic of this and other-worldliness.

Master/Disciple Relationship, Charisma, and Self-Cultivation

Our review of the Forest Movement and Ch'an/Zen will reveal affinities across numerous domains. At the heart of these trends is a cluster of forms that shape the soteriological impulse. Charismatic leaders and self-cultivation are the forms through which awakening will be pursued. Despite differences in certain practices, heritage, and cultural context, the shared commitment to direct knowing under the tutelage of awakened masters imparts a similar tone and tactics. This venture is accompanied by a critical stance toward normative Buddhism. The nexus of these factors: a soteriological telos channeled through realized masters and sought through personal cultivation generates the guiding dynamic of both movements.

The master/disciple bond is the lifeline of dhamma transmission and also the essential social relationship for the movement. If the forest wilderness and the sacred mountains are the habitat, it is the master/discipline that is its constitution. The forest movement begins not as a formal organization with a set membership and hierarchy but as a web of teacher/student relationships. Even third- and fourth-generation forest monks find their self-definition in their relationship to a teaching line. It is through *saai* (lineage) or *nikaya* (family) that membership, in/group/out group location, is determined. Monks within the same sub-lineage co-lead retreats. Each remains loyal to its teacher. While temples take president over teachers in the organizational structure of Thai Buddhism, in practice they are conflated. The monks of Ajahn Chah are the monks of Wat Pah Pong and its subsidiary temples.

Zen monks are known by with whom they practiced. From the mountains of China to the hills of California, teachers are identified by who granted their teaching certificate, *inka* (Jp,). For instance, during travels in China and Taiwan, the researcher is not asked for his temple affiliation, but the teacher one studies with. Questionable transmission leads to dispute on legitimacy of some Western teachers. If anything, the nascent condition of Western Zen and the absence of national or international organizational direction have made the teacher link all the more important. Where formal tradition is weak, charisma is paramount.

The shaping function of the teaching network applies to both the contemporary Thai Movement and the modern Sri Lankan movement. The forest movement in general mimics the earliest sangha organization of ancient India. Carrithers (1983, p. 142) characterizes the primitive sangha as composed of "little cells" spread across northern India functioning with considerable autonomy. Forest monks also typically wandered in small groups of fellow practitioners with the key difference being a relationship with a charismatic leader. All *saai* and *nikaya* trace themselves to charismatic figure whose descendants have been characterized as pupillaries, communities of pupils, with allegiance to the same teacher. For example, the nearly 80 satellite monasteries of Ajahn Chah are a "pupillary network" that constitutes a sub-lineage within the Forest Movement (Taylor 1993, p. 125).

Here, the term charisma does not fully conform to Weber's ideal type. While some masters may come close to being autonomous sources of authority (as was the Buddha), typically other teachers and even scripture provide counter weights. Ajahn Mun, for example, asserted that scripture could only correctly interpreted by those who had directly attained Dhamma. While not seen as the original fount of wisdom, the teacher is the living embodiment of Dhamma and it is only through relationship with him that guidance and verification of attainment can be assured. The charismatic status of the teacher is the elementary and essential social relationship of the Forest Movement.

The story of Ch'an/Zen is in part a tale of seekers in search of a master. Hiuke stands in the snow before Bodhidharma. Joshu spends the better part of his 120 years searching the Middle Kingdom for teachers to perfect his understanding. Dogen and Bankie exhaust the teachers of Japan. Students are dismissed and sent to other masters. The life of the Zen student revolves around the master. Even with the emergence of monasteries, the master/disciple relationship remained the vital thread. Community forms out of the shared status as students of a given teacher. The role of the Zen master not only reaches backwards in time to the Buddha but also spreads out horizontally in a web to form contemporary community. In Buddhist terms, the teacher generates sangha, community. This pattern of teacher-based community replicates what we know of primitive Buddhism and the Buddha's sangha making function. Here, again charismatic Buddhism replicates a dynamic of early Buddhism. Seekers do not join an organization; they find a teacher. At no point in the researcher's involvement with Ch'an/Zen did students seek verification of their practice via texts. The teacher, as the Buddha's representative, was the arbiter of truth.

In addition to the charismatic nature of the teacher/student relationship, there is an additional component that decisively shapes the forest and Ch'an/Zen's organizational web. The soteriological project of Buddhism is not based on sacrament or worship but demands self-cultivation. Although charismatic monks of the forest were later venerated by the lay public through relics, amulets, or merit making rites, the true fruit or *phon* (Th.) of the movement was self-liberation and this required self-cultivation through practice, or *patibat* (Th.). In turn, practice demanded direct instruction from the master. Carrithers pinpoints the implication of this relationship for the social world of the forest monk.

> *[But] the importance of individual self-cultivation has another significance for the Sangha organization as well: that organization must be small-scale, carried out within a face to face community.* (1983, p. 143)

Zen is also built around a soteriological project molded by charismatic teacher/student relationships and a demand for self-cultivation. Students were often heard to speak of this or that teacher's Zen, indicating the fusion of teaching and practice with the teacher. The contemporary phrasing has historical roots in Ch'an's internal sectarianism. Northern School vs. the Southern School, the five Houses or branches of Ch'an are all identified with seminal personalities. The survivors of these rivalries, today's Soto and Rinzai sects, bear the names of their supposed founding ancestors. Dongshan and Caoshan are contracted into Caodong (Ch.) or Soto (Jp.), and of course Linchi being posthumously awarded the sect named after him, Rinzai. The ultimate indicator of the centrality of the charismatic teaching relationship is demonstrated by the mandatory face-to-face encounter which is the hallmark of Ch'an/Zen training.

Together, the teacher/student relationship, charismatic authority, and self-cultivation constitute a complex that functions to generate the inner dynamic of Ch'an/Zen and the Forest Tradition. It is to the actual operation of these key relationships and its consequent teaching tactics that we now turn.

Summary

The history of Ch'an/Zen and the Thai Forest Tradition was traced from the respective genesis stories of Bodhidharma and Ajahn Mun. In the case of the forest monks as a general social movement, their existence as a recognized community as distinct from the pariyat or scholar monks of the city extended back to the earliest days of the sangha. Each movement was characterized by a tension with mainstream, urban-based Buddhism which they countered with their forest and mountain preferences and meditation-based practice. In contrast to scholasticism and a distant Nibbana, forest and mountain monks embrace a soteriological project of personally experienced liberation. Early Ch'an was found to hold Imperial authority at arm's length relishing its tales of defiance. This iconoclastic stance became a central theme in Ch'an/Zen's self-advertisement. Similarly, the Thai Forest Movement and the figures associated with Ajahn Mun were seen to be balancing their royal affiliations and their suspicion of the official sangha. The Forest Tradition and Ch'an/Zen carried, at least in their initial phase, the other-worldly polarity of Buddhism's dialectical tension between an other-worldly telos and an everyday accommodation. The interplay of charisma and self-cultivation as necessary for liberation gave social form to the soteriological project and in turn shaped Ch'an/Zen and the Thai Forest Movement's activity, style, and intellectual tone. That both schools are constituted by a similar complex of charismatic lineage, self-cultivation, and face-to-face instruction in the service of individual liberation suggests an explanation of their independently derived similarities. This working hypothesis invites elaboration and a close examination the actual teaching tactics and interaction patterns of the teacher/disciple relationship.

References

Burger, E. (Producer/Director). (2007). *Amongst the white clouds* [Video]. USA: Independent.

Carrithers, M. (1983). *The forest monks of Sri Lanka*. Delhi: Oxford University Press.

Ferguson, A. (2000). *Zen's Chinese heritage*. Boston, MA: Wisdom.
Ferguson, A. (2012). *Tracking Bodhidharma*. Berkeley, CA: Counterpoint Press. Kindle edition.
Ling, T. (1973). *The Buddha: Buddhist civilization in India and Ceylon*. London: Temple Smith.
Mah Bua. (1995a). The Venerable Phra Acariya Mun, Samudra Sakorn Wat Prajayarangsi.
McDaniel, J. (2011). *The lovelorn ghost and the magical monk*. New York: Columbia University Press.
Porter, B. (1993). *Road to heaven: Encounter with Chinese hermits*. Berkeley, CA: Counterpoint.
Red Pine. (1987). *Zen teaching of Bodhidharma*. New York: North Point Press.
Schumann, H. W. (2004). *The historical Buddha*. New York: Penguin Group.
Spiro, M. (1970). *Buddhism and society: A great tradition*. New York: Harper and Row.
Suzuki, D. T. (1978). *The Lankavatara sutra*. New York: Grove Press.
Tambiah, S. (1976). *World conqueror and World renouncer*. Cambridge: Cambridge University Press.
Taylor, J. L. (1993). *Forest monks and the nation-state*. Singapore: Institute of Southeast Asian Studies.
Tiyavanich, K. (1997). *Forest recollections*. Chiang Mai: Silkworm Books.
Watt, D. (1984). *Thailand: A short history*. New Haven, CT: Yale University Press.
Weber, M. (1993). *The sociology of religion*. Boston: Beacon Pr.

CHAPTER 3

Fish Sauce and Plums: Teaching Tactics

Introduction

Central to the identity, legitimacy, and organization of both the Thai Forest Movement and Ch'an/Zen is the teacher/ student relationship. This chapter will examine their teaching tactics and style and consider the pivotal role of the teacher/student relationship in both movements. The history of Ch'an/Zen is largely written in the teaching stories of master and disciple. Certainly there is an institutional history which has been documented. Ch'an's movement from China to Korea, Japan and Vietnam, and now to the West is a major chapter in world religion and culture (Dumoulin 2005a, b). Zen's impact on its larger culture, especially the arts, has attracted both scholarly and popular treatments (Hoover 1977). Nevertheless, it is the unorthodox interplay of master and student, and the transmission of wisdom from one to the other, that are its backbone and signature. The communicative acts that constitute the teacher/student relationship are both Ch'an/Zen's inner dialectic and its self-presentation to the world. What is Zen? Leaving aside enigmatic allusions to an ineffable absolute, the answer would no doubt be the history, not of any large institution, but of the teaching relationships and the tactics of awakening that compose a stream of transmitted Dharma.

The centrality of the teaching relationship and its teaching events to identity and legitimacy is equally true for the Thai Forest Tradition which, unlike the Ch'an school, is not an explicit sect. There are no membership

A.R. Lopez, *Buddhist Revivalist Movements*,
DOI 10.1057/978-1-137-54086-7_3

cards or special insignia, and while monks of the tradition follow a strict regimen, this code is neither an exclusive nor an alternate Vinaya. As Thammayut monks, their dress and behavior do subtly set them apart from the Mahanikai, the older and larger denomination in Thai Buddhism, yet forest monks are outwardly indistinguishable from their urban centered Thammayut brethren. Furthermore, some of the Forest monks including prominent members are Mahanikai. In short, official forms do not designate identity. Being a forest monk, therefore, is a matter of teacher/student lineage. With whom does one study/practice? The answer to this question determines membership. An "oral lineage" consisting of face-to-face communication is preeminent over formal organization.

Thread and Transmission

As a self-conscious movement, the thread or *sen* (Th.) which binds the Thai Forest Movement is the student teacher bond. Fading photos and fanciful group portraits present a school defined by wizened masters sitting together, usually with Ajahn Mun at the apex or in the foreground. Like a graduation photo, it conveys that these monks are classmates in a distinct school of Dhamma. This depiction is confirmed by casual conversation. Students and teachers continually reference one another, usually with utmost respect. Disciples both ordained and lay swap stories and practice recommendations for *wats* and *ajahns*. Without formal organization, informal conversation conveys a strong in-group identity secured to a web of forest masters.

The Ch'an/Zen obsession with lineage and transmission through lineage is itself legendary. A list of transmission masters, however historically dubious, extending back to the Buddha is chanted daily. Major sub-sects within Ch'an are named after their purported founders. As already noted, Caodong (Soto, Jp.) being a conflation of the names Dongshan and Caoshan, two T'ang dynasty teachers. Linchi (Rinzai, Jp.) sect takes its name from the teacher of the same name, also a T'ang dynasty figure. Although modern scholarship has cast doubt on, if not shredded, Chan/Zen lineage charts especially those attempting to reach back into India, their continued assertion confirms the decisive role of lineage in Ch'an/Zen's self-conception (McRae 2003; Wu 2008).

For the Zen sect, teacher to student transmission is critical mythology for legitimacy of realization. How does one know that this knowledge is true? When the methods are unorthodox and there is no reliance on scripture on what does one rely? The Ch'an/Zen answer is to point to an imagined unbroken line of patriarchs and teachers winding back through history to Shakyamuni Buddha himself. The Buddha silently raises a flower and Mahakasyapa smiles, presumably knowing the ultimate. The Dharma has been transmitted. That the tale is probably fancy and not historical fact only affirms the vital role of person to person, teacher to student transmission in the Zen. This transaction may be without words and is termed "mind to mind transmission." The communications which constitute the transmission are at the heart of Ch'an/Zen.

The importance of lineage as a line of inquiry and as an organizing social dynamic has also recently been noted by scholars of South East Asian Buddhism. What emerges from this perspective is not the Buddhism of organizational charts and catechisms compiled in ecclesiastic offices, but a dhamma practiced by its enthusiasts, ordained and lay. The activity is fluid and intensely personal and entails a bonding with "your monk." Actors are seen as having a "repertoire" of religious behaviors that are collected and activated over a life time (McDaniel 2011, p. 9). Sensitivity to this dimension of religious life is well suited to the study of the Thai Forest Tradition. The texture of the teacher/student relationship is woven by the weave of the tactics of instruction. These distinctive communication patterns reveal the "living dhamma" of the Thai Tradition and its affinity with Ch'an/Zen.

Telos and Charisma

The teaching tactics of both traditions need to be understood within their broader context defined by the telos or the ultimate purpose of the school, its teachers, and students. Specific teaching tactics occur within the framework of their meta-purpose which also determines their function and evaluation. No matter how varied and extreme these linguistic and behavioral maneuvers, the goal and evaluative standard are singular. It has been labeled as pragmatic-soteriological (Wang 2003, p. 11). The teaching is pragmatic in that it is judged in terms of its consequences for the student rather than in terms of abstract religious principles. One looks in vain to find tactics characterized by conventional understandings of Buddhist values: generosity, loving-kindness, or equanimity. Instead, communications are judged by their effect on the student. Does it, to use the words of the Buddha himself, tend toward liberation.

The teaching is soteriological in that the goal is the disciple's Awakening. Progress on the spiritual path, ideally liberation in this life time, is the raison d'être of the relationship. While the ultimate goal is conceptualized somewhat differently in Theravada and Mahayana, they agree in that the guiding value is freedom from the limitations of the human condition. This is the tacit commitment of teacher and disciple. Teachers are inevitably involved in other kinds of relationships, especially with lay persons, seeking merit, and an improved material circumstance. Advice on worldly matters may be sought and cows brought for blessing. Such requests may be entertained or summarily dismissed, but they are clearly of a different order from the intention that guides the master/disciple relationship. Here, the commitment is to enlightenment and nothing less.

The lives of the Thai forest masters portray an uncompromising commitment to a soteriological goal. Placing one's life at risk to wild animals, disease, malnutrition, and the elements is a regular theme in their biographies and autobiographies (Tiyavanich 1997, p. 79). A common attitude is that death is unavoidable; therefore, there is no reason to compromise the quest in order to delay the inevitable. If you are sick, take medicine and lie down. If you recover, fine; if not, so it goes. If the tiger eats you or the wild elephant tramples you, resume your journey in the next life. Presumably, there were monks who perished alone in the forest. The risks were real and not merely the hyperbole of hagiography. The willingness to take such chances and endure hardships affirms the unsurpassed value being sought.

Zen stories abound of disciples driven to their mental and physical limits often climaxing in transcendent breakthroughs. Bankei, a fierce Zen monk of sixteenth-century Japan, resigned himself to a death probably by "monks' disease," tuberculosis. Watching bloody sputum he had coughed up slide down the wall of his hermitage, he awakened to the Unborn Buddha Mind (Waddell 2000, p. 10). A modern Korean Zen (Son) master lived through her country's hard winter without shelter, burying herself in the ground (Daehaeng 2007, p. lox). A Chinese Ch'an monk informant described a three month retreat of sitting upright to sleep, hurrying through meager meals, and suffering the winter elements. "You would die," he predicted. I concurred. Why do this? To attain the freedom the Buddha knew is the answer.

The pragmatic-soteriological project is played out within a charismatic relationship. The master is understood as having personally realized liberation which affords him a credit not granted other monks or dhamma practitioners, and, as we shall see, puts scripture in the shade. Prior to acceptance by a kruba ajahn or sifu/roshi, students report prophetic dreams or having undergone physical challenges to access the teacher. For example, a "psychic intimacy" is present in Ajahn Khao's story. After

Ajahn, Mun's death would appear in Khao Analayo's meditation, both day and night, to offer answers to practice questions (Maha Bua 2006, p. 104). Maha Bua, the great biographer of the Forest Movement, comments on his initial meetings with Ajahn Mun, and his immediate trust. "It is not an exaggeration," Maha Bua assures us, "that his disciples had complete faith and reverence in him" (1995a, p. 259). Trust and faith amplify any evaluation of sheer competence to expand the relationship into a globalized charismatic connection.

Ch'an stories often recount students being initially rebuffed or forced to undergo extreme physical and mental trials. Even today, monks seeking entrance to monasteries of notable teachers will be forced to wait outside enduring the elements for some days before being granted admission. Such tests along with the deference displayed by other disciples, recommendation by other senior teacher/monks, and circulating reports of the teacher's arduous practice and meditative attainments, construct a person of towering authority, a veritable litmus test of living Dhammic Truth.

Teachers in these traditions, of course, have their titles. In the Ch'an tradition, "*sifu*" generally indicates a master teacher, and is the honorific generally used by disciples rather the personal or dharmic-monk name. In the Japanese, an old venerable teacher is "*roshi*," while a younger teacher is "*sensei*." The term sensei can be found outside the Buddhist tradition to indicate a teacher any number of arts and academic endeavors. The Thai tradition offers a rich lexicon of respectful titles. The most affectionate and yet honorific is "*luangpor*," or venerable father. The use of the kinship term articulates a relationship of super-subordination and veneration. The terms "*kruba*" and "*ajahn*" are also employed, sometimes together. The "*kruba ajahns*" of the forest tradition is a common designation. The term ajahn, a teacher who has an advanced degree, is used in secular circumstances such as with any school teacher possessed of a higher degree. "Kruba," like the word "ajahn," is not reserved for teachers of the forest school. It is a title granted to respected monks in Thai Theravada Buddhism. Kruba Srivichai, the patron saint of Lanna, is known by this designation, but was not of the forest lineage. It is one of long list of honorifics that are not ecclesiastical offices. While the Forest Tradition does not monopolize these titles, what is of note is that they confirm a hierarchy of respect outside the formal sangha titles.

The masters of both traditions are the ultimate arbiters of doctrine. Of course, this is the essential component of Weber's charismatic ideal-type (Weber 1946, p. 245). The charismatic master has been lit up by his attainment of the sacred. While scripture and commentary have a role and receive some study and may be quoted to buttress oral teaching, the final authority is the teacher. The teacher guides and verifies the student's progress. The teacher gives final interpretation to religious experience and invents his own labels and understandings which may differ from classical dogma. Charisma, therefore, is inherently creative and keeps tradition alive, but also entails stepping out on a high wire without the safety net of impersonal institutional authority. Together, the pragmatic-soteriological project and charismatic authority constitute the stage setting of the teaching tactics. Here, the operating principle is that all is fair in the war on suffering and ignorance.

Teaching Contexts/Teaching Moments

The canopy of pragmatic-soteriological project and its tent pole of charismatic authority cover a wide range of social contexts. Teaching moments are everywhere. The settings range from public gatherings to private, one-on-one meeting. Teaching moments maybe initiated by the master requesting the student "report in" or they may follow student requests for an interview. The Ch'an/Zen school is notable for its formalized teacher/student one-on-one meetings (*sansen, dokusan,* Jp.) and for ritualized public dharma combat between sangha and master (*shoshan* Jp.). These encounters occur frequently during meditation retreats and as a regular aspect of temple life. However, there are also spontaneous encounters. The researcher has witnessed contemporary Ch'an master Sheng Yen on a casual walk transforming a student's observation about cow dung in a field into a koan-like meditation question, or Eido Roshi, a teacher in the New York area, suddenly asking a new student "Where do you come from?" The question, of course, is existential and not a request for a zip code.

The Thai Forest Tradition does not have the highly ritualized encounter of Ch'an/Zen. Meetings, however, are both private and public, requested and demanded, scheduled and spontaneous. Ajahn Chah, for example, frequently used public meetings to single out monks for pointed feedback. He would recount incidents or draw exuberant analogies to expose foibles, much to the mirth of his audience. "I really like to fool around," Chah confesses (Breiter 2004, loc. 1107). Ajahn Maha Bua is recorded as even dismissing a recalcitrant student at an open gathering (Silaratano 2009, p. 159). At the other end of the public–private spectrum, written reminiscences and oral informants document casual teaching moments as when the teacher was strolling by their *kuti*, meditation hut, or in a chance encounter. Ajahn Lee recounts being taught meditation by Ajahn Mun while on *pindabat*, alms rounds (Lee 1991, pp. 35–37). In Lee's memoirs, Ajahn Mun is depicted as both skillful and persistent in his use of both the social and natural environments for dhamma lessons. Chance encounters were also teaching opportunities. A monk reports wrestling with a private obsession in his *kuti*, when his teacher passed by and directly addressed the content of his mind. Such incidents enhance charismatic authority but also illustrate that teaching can and does occur anytime and anywhere, so be awake!

Historical Antecedents

The teaching tactics cover a wide range of immediate situations, but they have a background in Buddhist history and classical thought, that affords them meaning and justification. The radical nature of the teaching style ironically finds precedent in classical Buddhism. The emphasis on personal experience, direct knowing, the quest for a final resolution, and the pragmatic approach are characteristics of Buddhism since its earliest articulation. The root texts, as we have already noted, define the Dhamma as *akalito* to be known in this very life, and *ehipisakka*, inviting one to see for one's self. Personal experience is primary. This is the working premise of both the Forest and Ch'an schools. Equally, the undeniable pragmatic nature of the Buddha's teaching provides further legitimacy and inspiration for the Ch'an and Forest projects. The Buddha's managing of young Brahmins who come looking for God is illustrative. After suggesting that those who have advised them have done so without any direct knowing, the Awakened One elicits their agreement that to be in the realm of God, Brahma, they would be infused with divine qualities such as loving-kindness. The Buddha then teaches a practice to evoke *metta* (Wallis 2007, p. 9). He does not challenge their faith; rather he asks them to operationalize it and then offers an activity to actualize it.

The Buddha is the mega-charismatic of the tradition. While scripture speaks of other Buddhas preceding the Buddha of this historical epoch, the man who walked northern India 500 years before the Common Era (C.E.) is recognized as "self-arisen, self-taught." The Buddha was the dispenser of truth, the arbiter of doctrine, and the functional dictator of the sangha. From the loftiest of teachings to the micro-regulations of the Code of Conduct, the Vinaya, the Buddha's word was incontrovertible truth and law. While no Buddhist teacher has claimed such a status, the charismatic style of authority has been replicated even if at a more modest level.

The teaching methods, while going beyond the Buddha's, are nonetheless presaged by stories recounted in the earliest texts. Along with the doctrinal emphasis on direct knowing and the charismatic teacher, the Buddha's ample use of paradox and allegory anticipates the language of these movements. The Buddha was a first rate storyteller. Compilations of the Master's parables are in print (Burlingame 1999). Ch'an and Forest teaching tactics can be viewed as applications and extensions to concrete practice situations of general Buddhist principles and teaching style. They do not arise *ex nihilo*, but extend the spirit of both doctrine and teaching encounters recorded in the Buddhist Canon. Ch'an and the Forest Tradition self-consciously claim to be reviving and maintaining the spirit of the Buddha's Way. Despite being innovative and distinctive, these movements still have a claim on the Master's robe and bowl.

Aside from any self-serving sectarian arguments, both Ch'an/Zen and the Forest Tradition qualify as "revivalist" Buddhism. They bring into the foreground legitimate aspects of Buddhist practice/teaching that are often marginalized by a state run Buddhism primarily concerned with buttressing social hierarchy and dolling out consolation. Unlike conservative revivalism that attempts to resuscitate or reinforce traditional social forms, these movements unsheathe the cutting edge of Buddhadharma that is often blunted by community/state Buddhism. The radical, indeed revolutionary, character of these radical revivalists is dramatically displayed in their teaching tactics.

Ch'an and Forest Teaching Tactics

Any exploration of the teaching tactics of these two movements should be balanced with recognition of their use conventional modes of imparting Dhamma. Both Ch'an /Zen and the Forest Tradition use the classical sermon format that is shared with others in the Mahayana and Theravada schools. Zen Masters offer commentary on *koans, teisho* (Jp.), and within the Soto branch roshis will typically offer during the course of their careers a series of talks on a classic text such as Shitou's *Unity of Sameness and Difference*. Likewise, Thai teachers also employ a more traditional format, "the dhamma talk," *dasana*. However, the more traditional teaching formats ultimately contrast with and bring into focus what is the distinctive dynamic of the teacher/disciple communication. The teaching tactics are both the substance and style that makes these movements unique and yet startlingly similar to one another.

Ch'an/Zen teaching tales have long intrigued, bewildered, and inspired both East and West. For the public, they may be seen a variety Asian absurdist theater or conundrums to be bantered about at cocktail parties or in the halls of academia. For serious students within the tradition, recorded teaching episodes are meditation themes whose penetration entails transcendental wisdom. More recently teaching stories of the Thai Forest Movement have become available through autobiographies, biographies, and reminiscences, and via participant observation accounts. The interplay, both verbal and non-verbal, is where, as one my informants suggested, "the Dhamma is alive." Here in these exchanges between teacher and disciple, in these micro-interactions, we find the life blood of both movements.

The following tactics are common to both movements, indicating their close affinity even at the level of micro-interaction. The tactics are sorted by descriptive type and do not mutually exclude one another. Any given communicative act may be analyzed into more than a single tactic. As there is no handbook of tactics, they are employed according to the teacher's style and the situation. Nor is there a requirement or quota on their use. Teachers vary in the frequency and type of tactic used.

Utilization

Utilization refers to the incorporation and use of immediate events and the environment as a teaching device. The term is employed in both education and hypnotherapy where it points toward a spontaneous, unrehearsed use of the moment (Erickson 1996, p. 67). A corollary to utilization is the de-emphasis of book learning. For example, referring to walks with Ajahn Mun's, Ajahn Lee says: "No matter what we passed—houses or roads—he'd always make it an object lesson." An attractive woman was an occasion to examine the mind and perform an *ashuba* meditation, analysis of the body. This tactic of utilization leads Ajahn Lee to unequivocally declare, "Nature is the teacher." "I was no longer worried about studying the scriptures"(Lee 1991, p. 161). Lee's conclusion echoes the Zen dictum of a teaching not relying on the scriptures. Instead there is a direct pointing at the immediate environment.

Zen stories abound with incidences of the environment teaching. Kyogen awakens at the sound of a stone striking bamboo. Tung-shan awakens upon seeing his image in a stream. "Everywhere I look it is me!" (Powell 1986, p. 28). Dogen, the thirteenth-century transmitter of Ch'an to Japan, hears the one true thing, "black rain on Fukakusa temple" (Stryk 1994, p. 2). Utilization is considered a creative tactic in education, yet in Buddhist training it assumes an additional value: The mind is turned to the here and now. The mind of speculation and thought is replaced by perception of the immediate. Utilization employs and actualizes the Buddhist teaching that here and now is the zone of awakening. The technique is innovative, but it contains a classic Buddhist principle.

Paradoxical Speech

Paradox not only disrupts and baffles conventional thought but also demands resolution from the recipient. "You have to say something!" as a contemporary Zen Master, Katagiri Roshi put it (Katagiri 2000). Ajahn Chah (2005) particularly favored this tactic. "You have seen flowing water?" Yes. "You have seen still water." Yes. "But have you seen still flowing water?" (p. 104) To a monk requesting discipleship, Chah replies: "If you seek a teacher you won't find a teacher. If you have a teacher you have no teacher. If you stay with me you won't see me. If you give up the teacher you will find the teacher" (p. 141) After a sleepless night struggling with Chah's paradoxes, the monk presents his understanding. Chah approves. The monk departs. The need to be a disciple has been dissolved in a new perspective on path and teacher. On other occasions, Chah would pose the conundrum of what one could do if you could not move to the side, forward or back, up or down, yet could not stay still! (Brahm, http://www.dhamaloka.org.au). This Zen-like command placed the listener in an impossible situation that could only be resolved through a non-rational response.

The Zen tradition is famous for its puzzling anecdotes and paradoxical commands. Mazu orders his disciple Layman Pang to swallow the West River in a single gulp (Ferguson 2000, p. 94). A classic Zen tale has a master in response to a monk's dumbfounded silence in the face of paradox shouting, "Speak, speak!" (Ferguson 2000, p. 259). What is critical is that he is asked to do something. Move. Swallow. Speak. You must say something. The paradox is an explicit or implied command for action.

It is not only one's cognitive map that is crumpled but one's "vocabulary of motives" (Mills 1940) is tossed out the window. How can one act on this new terrain? How can Layman P'ang swallow the West River? (And in a single gulp to boot!) Ajahn Chah offers some penetrating insights into this tactic. Alone among the early forest teachers, no doubt through his extensive contact with Western monks and translated texts, he became aware of Ch'an/Zen and their common teaching ploys. Paradoxical command forced the student to respond out of an alternate (transcendent) wisdom. Action has to emerge unfiltered by the mind's discriminating mesh. Chah believed that he and the Ch'an masters of T'ang China were up to the same game (Breiter 2004). Paradox, both at the level of cognition and behavior, is a potent tool in the teaching repertoire of the masters of mountains and the forests.

Indeterminate/Indirect Speech and Language

Indeterminate language and speech consist of enigmatic expressions, unconventional word usage, riddles, and allegory which also compel the listener to actively engage. What does it all mean? One day out of the blue, Ajahn Fuang asked, "If your clothing fell down into a cesspool would you pick it up?" Practical answers were dismissed (Fuang 1999, p. 19). Ajahn Chah's mentor, Ajahn Kinnaree, used bizarre expressions that left the young Chah and his fellow monks bewildered. Whenever Ajahn Kinnaree left the temple, he would shout, "Don't let the dogs shit in the temple." But there were no dogs! Only subsequently did they realize he was referring to the undisciplined minds of monks (Chah 2005, p. xx).

Aspects of indeterminate language are to be found in the Buddha's teaching. In the famous episode between the Buddha and Angulimala, the serial killer in pursuing his intended victim fails to close the distance between himself and the calmly walking Buddha. Frustrated he commands, "Stop!" "I have already stopped. When will you stop?" (Thanissaro 2003). The Buddha plays on multiple possible referents to the word, "stop." There is little wonder that this play with words would re-appear in later Buddhist movements.

Both schools extensively employ allegorical tales often featuring a menagerie of foxes, hares, tortoises that teach graphically yet indirectly. Whether part of a classic composition, adapted from folk tales, or generated on the spot, they fulfill the Zen adage, "never say too plainly" (Wang 2003, p. 161). Why not? The medium of the message is also a teaching.

The linear, discursive mind is frustrated while more associative cognitive processes are insisted upon. A different way of knowing is being taught not simply a new content.

Indirect language, the language of allusion and indeterminate meaning, takes numerous forms. Ajahn Mun was a player of the Isan art of Morlam, a story telling contest that relied on spontaneous rhyming, word play, and wit. Speech can be combative and clever. The similarity to Zen both in its indirect language and in its testing style is evident. Use of this skill, it has been suggested, carried over to Mun's reportedly effective teaching delivery. Bhikkhu Thanissaro observes, that "indirectness of such a style could make it very suggestive, giving it a direct impact on a subliminal level" (Taylor 1993, p. 76).

However, an account of indirect language would not be complete without referencing the role of poetry in Ch'an/Zen literature and the *kung an* (Ch.) or *koan* (Jp.). While these literary forms warrant and have received in-depth study, we can note that they are Ch'an's expeditions into the inexpressible, returning with enigmatic and evocative words. The poetry of Ch'an/Zen can take more expository and philosophic forms and often did among the early Ancestors, nevertheless, it evolved into shorter a more imagistic, suggestive pattern. Hongzhi, dubbed the poet of enlightenment, offers a vision of the awakened mind:

> The water is clear right down to the bottom, fish lazily swim on.
> The sky is vast without end, birds fly far into the distance. (Leighton 2000, p. 61)

Ryokan, Zen poet monk tells us:

> Next to my hermitage there is an ancient bamboo grove;
> Never changing it awaits my return.

(Stevens 1977, p. 46)

The Zen *koan* is perhaps the ultimate example of forced engagement of indeterminate language. The koan, literally a public case record of a classic teaching incident, functions as an assigned meditation subject that the student must resolve. By arousing "Great Doubt" and "sweating white beads," the disciple forces a breakthrough to be approved (or rejected) by the master. While allegories, poems, and parables tease the mind into involvement, the koan seeks to grab the student by the throat.

The koan also calls on poetry, as in its classical use, its resolution was literally "capped" by the disciple's poem.

Perhaps the quintessential indeterminate communication and paradoxically the most direct expression is the Ch'an shout or *kwatz*. Records indicate the Ch'an masters would use the shout often in response to logically framed inquires. The sound or kwatz has no meaning; it is not a known word. This audio blast renders irrelevant language-based meanings of conventional correspondences. The truth is beyond words, yet it immediately manifests. Ultimate reality explodes out of the teacher's mouth. Reality is an actualization/expression not a detached reflection. The forest tradition, while not devoid of poetic production, has nothing to correspond to Ch'an's kwatz, or the poetry of pure, immediate perception stripped of all abstraction. In these domains, Ch'an is unmatched.

Pith Instruction

Along with indeterminate and indirect speech, teachers use what can be called abbreviated or pith talk. The teaching slogan may be "say only the necessary" (Wang 2003, p. 161). This tactic replicates other teaching gestures. Communication is to the point of practice not the elaboration of theory. By saying only the necessary, like never saying too plainly, there is also ample room for creative engagement rather than rote learning. The disciple is thrown back on their own resources, yet with some guidance. Even Ajahn Mun for instance was largely left to his own devices in taming his restless mind (Maha Bua 1995b, p. 16). Pith instruction offers clarity, yet on further reflection, they may be indeterminate as well. For instance, one is told to uproot defilements, a common directive in forest circles. Yet when inquiring as to exactly how one does this, one is told to find out for oneself.

Zen and the forest tradition as they address a wider and often lay audience have made practice instructions more explicit. For example, Sambo Kyodan, a lay centered synthesis of Soto and Rinzai, provides explicit instruction on koan work. This contrasts with the more classic approach of hurling the koan at the student and letting them discover how to work the conundrum. Likewise with large organized retreats, the forest tradition has systematized *patibat*, or practice. Nevertheless, pith instruction and their directness and accompanying indeterminate nature continue to be used especially with more advanced practitioners. Frequently, the researcher was challenged to arouse his own panna, wisdom, to resolve a practice question.

Perhaps the quintessential pith instruction coming from Ajahn Chah is "Let go." In answer to the overly complex, convoluted mind of the disciple was "Let go." What is Buddhism? What is practice? What are the precepts? "Let go" (Chah 2002, p. 25). In effect, Chah tells his students to forget about abstract dhamma principles and just keep letting go. That is Buddhism. Chah apparently learned this mode of pith instruction from his teacher, Ajahn Thongrat. "Drill the hole right in line with the dowel," he instructs. "If it comes low, jump over its head. If it comes high, slip under it." The young Chah is bewildered. Only later in his practice does he grasp that this is a pith instruction on handling defilements (Chah 2013, p. 127).

Modeling Behavior

An axiom of socialization theory is that we learn by from the "what and how" of the actions of the significant other. We catch on through observation and taking the point of view of the other. Upon sharing this with a monk who has practiced with several well-known forest teachers and is now himself an abbot, he enthusiastically agreed. In observing his present master, he noted, "He does everything very quickly, but very precisely." He moved his hands in rapid chopping gesture and suggested that this behavior conveyed continuous, sharp mindfulness. The teacher's behavior was an action model of a valued inner state. By imagining occupying the other's body or replicating the body language, the student takes on a feel for a state that can only be imperfectly rendered in linear language. Another forest monk observes that the teaching is less about understanding and more about training (Breiter 2004, loc. 368). One learns by doing. One does by observing how the other does. A Ch'an monk who practiced in Thailand at forest temples called keeping the precepts, "samadhi itself." Conforming to a pattern of behavior imparts a mental set.

The Soto Zen practice slogan, "The body leads, the mind follows," articulates this teaching principle. Perhaps an extreme in modeling is reported by contemporary Zen teacher, Jakusho Kwong. He spent an entire retreat synchronizing with his teacher's breathing. The teaching method might be phrased as, do as I do and do not think (too much) about what I say. What Kwong reports learning from the retreat are not words of wisdom, but a breathing pattern, a way of being (Kwong 2007).

Modeling behavior is a maneuver that by-passes abstract learning. When being taught a task or in answer to a question about location, the task was invariably demonstrated or I was physically led to the destination. Upon being told to sweep leaves (an endless task in Thailand), the senior monk would inevitably demonstrate. Symbolic learning was kept to a minimum. Modeling, therefore, not only conveys content but a way of functioning beyond the conceptualizing mind. Residing at a Ch'an center or Forest Monastery entails a shift into a mode of observing and doing not thinking.

Confrontation

The term "confrontation" takes its technical meaning from psychoanalytic literature. The patient is challenged about behaviors including motives and consequences that he may ignore or rationalize. Upon visiting a Zen teacher with whom I was working, I was asked why I had come. My rather formless response drew a sharp question. "Don't you have enough trouble in your life?" Ajahn Chah was known to greet newcomers with a question. "Have you come here to die?" Teachers in both these traditions are trouble. In contrast to kind, smiling "grandmotherly" love, they offer tough love.

Several Thai Forest teachers have reputation for unvarnished confrontation. I was warned about visiting Ajahn Sa-ngop, an enlightened student of Mah Bua. He points out defilements very directly, I was told with a stabbing gesture toward the center of my chest. "Sometimes people don't come back." A monk came to Ajahn Chah and grandly told him of the exalted state he had achieved. "Well that's a little better than a dog." Chah retorted tuning his back and walking off (Chah 2005, p. 94). Ajahn Thate reports his teacher Ajahn Singh "really shook me up" upon challenging Thate's obstinate opinions about his practice (Thate 1993, p. 82). Maha Bua has a reputation greeting visitors brusquely. Ajahn Khao Analayo reports his teacher's sharp feedback, "Ajaan Mun would still scold me, shredding me with his fierce language, which I reckon was right and suitable for someone like me who was always talking and so couldn't be quiet and contented. On the other hand, it was also quite beneficial because I was able to hear a Dhamma teaching that went straight to my heart" (Maha Bua 2006, pp. 54–55).

Confrontation is prominent in the Ch'an teaching style. As previously mentioned, Yumen visited a temple seeking entry. The teacher grabs him by the robe, demanding "Speak! Speak!" Upon hesitating, he is forcibly thrown out the front gate (Ferguson 2000, p. 259). A monk expounds on the emptiness of all phenomena only to have his nose violently twisted in a demonstration of his still all too solid flesh (Ferguson 2000, p. 77). False epiphanies are deflated, character traits are challenged, fixed views are attacked, and defilements are exposed through direct confrontation. While the forest tradition appears to primarily employ confrontation to turn the student from proceeding down false paths, the Ch'an application seems designed to force breakthroughs. Typically, Zen anecdotes featuring confrontation result in sudden awakenings.

Contradiction and the Deconstruction of the Sacred

Contradiction, the holding or switching between logically exclusive concepts is a tactic that pulls the conceptual rug out from underneath the student/disciple. Attachments to doctrine or views are targets to be challenged. The Buddha gives license to this enterprise in his comment that laypersons are attached to sense pleasure, but monks are attached to views. Caught in a thicket of views, he termed it in the *Aggi-Vacchagotta Sutta*. To get free of a thicket demands ruthless cutting.

Mazu, a towering figure in classic Ch'an, was famous for his teaching: Ordinary Mind is Buddha Mind. One day he sent one of his monks to visit a former student of his, now a teacher in his own right, with the message that he had a new teaching: Ordinary Mind is NOT Buddha Mind. The monk did as told and returned. What happened? Mazu asks. The monk reports the exchange of pleasantries and the inevitable inquires about health. Mazu hurries him on to the key point. What did he say about my new teaching? He just smiled and told me to tell you: "He says Ordinary Mind IS Buddha Mind." Upon hearing his old student's response, Mazu leapt to his feet exclaiming, "The plum is ripe!" His old student was "ripe" beyond words and philosophical positions. He was not relying on scripture or external authority. Presenting him with a contradictory teaching did not confuse him as it might have with an unripe plum. He knew directly (Wang 2003, p. 173). Intentional contradiction can unhinge the conceptually stuck student, but it can also be the test that confirms realization beyond words.

Contradiction can function as a pointed deconstruction of sacred categories. To deny Buddhist words is to deconstruct the sacred. Sacred constructions are still constructions to be jettisoned. They are not the Unconstructed that the Buddha speaks of in the *Udana Sutta*. Upon disembarking on the other shore, ditch the raft, the Buddha tells us. Ajahn

Chah on occasion used terms such as Original Mind and Pure Mind, yet on other occasions he denies them any validity. A monk, who was possibly enamored of these terms, appeals to Chah's prior teaching. Chah on this occasion brusquely dismisses his own teachings or at least their concepts. On another occasion, he is asked, "What is sotopanna (the first stage of Enlightenment)?" "Fish sauce," quips Chah. Ajahn Sumedho, Chah's foremost Western disciple, offers an explanation of his teacher's dismissal of a significant Buddhist idea, sotopanna. To the disturbed monk, Sumedho suggests that just as fish sauce gives taste to a meal so such concepts add flavor to the path, but do not take it too seriously (Breiter 2004, loc. 609). "Don't be an Arahant. Don't be a Buddha," Chah advises. The slogan has been posted on a wooden sign at Chah's international monastery. At the surface, we are being advised to leave the path and discard its goal. Upon deeper reflection, we are having our construction about the sacred dismantled.

Likewise Suzuki Roshi, the founder of San Francisco Zen Center, suggests that in the end, we are not even Buddhists. Self-conscious Buddhism is extra (Suzuki 1999, p. 123). Maha Bua may have used a similar teaching ploy. In an interview in the context of possible government action against Wat Dhammakai, he is asked is there or is there not a self? Both and neither he responds with a shy smile. He could be seen as just evading a political trap, but he may also have been undermining the attachment to either/or views. The disciple is forced back on the moment denuded of grand ideology and even everyday labels.

The student cannot hang on the literal language of Thai Forest Masters. To his students' confusion, Ajahn Thongrat insisted on mislabeling the gender of a water buffalo (Chah 2005, p. xxi). Apparently, a water buffalo by another label is still a water buffalo. Male or female from an ultimate point of view have no meaning. Ajahn Thiradhammo observed his teacher's inconsistent statements and contradictory advice. Only later did he realize they were flexible responses to situational requirements (Senior Sangha 2013, p. 8). Contradictory positions may both be wisdoms of the moment. Language is not allowed to fix perception or action.

Compare the above with the Buddha's announcing the approach of a Brahmin. The onlookers see only a commoner. Where upon the Buddha discourses on the true Brahmin being a state of the heart. His hook line, however, is speech that contradicts common sense. Forest and Zen teachers have apparently learned from the Master.

Compassionate Torment

The teaching is often intentionally aggravating. *Toramahn* is the Thai word for torture or torment and is openly acknowledged as a method among forest masters. "How was your stay with Ajahn Jun?" a monk is asked. Toramahn. "That's just his method of teaching," he is told. Paul Breiter (2004, loc. 1104), a former monk in the Thai tradition and Gou Gu (2012), a Ch'an monk, recounts being driven to the breaking point as attendants to their respective teachers. Ajahn Lee poked a hole in the wall to observe Ajahn Mun's way of arranging his quarters in order to stop being upbraided (Lee 1991, p. 37). Lee's successor, Ajahn Fuang, kept annoying people around the Wat to harass and test his disciples (Fuang 1999, p. 26). At a Bangkok talk, Ajahn Sumedho recounted being put to work breaking rocks for a new road, despite his protestations that he had come to meditate. Practice was being present regardless of the task, not just quiet meditation. The skillful persecution of disciples is a Zen art. Guo Gu, a dharma heir of master Sheng Yen, spent years as his attendant and like Ajahn Lee with Ajahn Mun was driven to beyond his limits. In response, he describes throwing himself into tasks and achieving a kind of samadhi of work (Gou Gu 2012). Perhaps what the old Chinese Master wanted.

The use of extreme and persistent "torture" has come in for questioning. Ajahns and senior monks at Wat Pah Pong, Ajahn Chah's primary monastery, have had doubts. The modern Ch'an Master Sheng Yen in the author's observation came to no longer employing the stick in the West. Instead, he invented new trials. On a meditation walk through the woods we come on an old Ch'an Master sitting by a stream, mindfully filling bowls to the brim with water. We receive our bowl and are told to take it to the Ch'an Hall. Spill a drop and you are banished from the Hall. The considerations are pragmatic not ethical. Frustrating the student's attachments by depriving them of sleep, assigning onerous jobs, and applying ruthless criticism and more is seen as a basic teaching tactic. Indeed, torture is an indication of the teacher's compassion which goes beyond soft "grandmotherly love." One student dubbed his Thai teacher, "The World's Most Compassionate Sadist" (Breiter 2004, loc. 324).

In the Zen setting, students report the one-on-one meeting with the teacher can be its own form of torture. After anxiously awaiting the chance to deliver the answer to one's meditation theme, the student is summarily dismissed with a ring of the bell before opening his mouth. There are reports of monks avoiding facing the master who are then carried then unwilling into the interview room and unceremoniously dumped before the teacher. Sheng Yen, who practiced in Japan as well, commented that the Japanese turn every inner barrier into a physical trial. One sits in the snow, is deprived of sleep, and eats quickly. When I asked Roshi Glassman what he thought of Rinzai temples in Japan, he had a one world answer, "Sadistic."

Intentional frustration is a tactic to be found in non-Buddhist settings as well. To transform the person is to frustrate their old ways of adjusting and defending themselves. Gestalt therapy, for example, defines the therapist's task as frustrating the client's maladaptive ploys as a way of promoting an explosion into new, and presumably authentic behavior (Perls 1969, p. 29). Similarly, the Buddhist tradition is interested in forcing a letting go of attachments. The disciple is pressured into letting go of sensual attachments (food, sleep, creature comforts) and psychological attachments to praise and a self-image of competence and likeability.

Function of the Tactics

Beyond a description of the forms of the tactics of instruction, what are their functional consequences? At the level of meta-purpose, of course, it is the attainment of awakening and liberation. Yet there are more immediate consequences that can be examined. The function/consequences of the tactics can grouped together under: shock, confusion, introducing new learning modalities, and forced non-attachment and engagement.

Shock and confusion are internal states which signal the collapse of old understandings and a condition of potentially creative chaos in which learning can occur. Shock is the surprise engendered by a behavior which defies expectations. Yumen is speechless when seized and commanded to speak. A monk is confused by Ajahn Chah's getting on all fours like a dog. Each suffers a disruption of old meanings. While the tactic may contain a hidden meaning, its mere capacity to undermine a once solid view of the world serves the pragmatic-soteriological project. Confusion can prepare the actor for the new. "Confusion" if viewed etymologically (from the old French) is to pour together. Therefore, confusion implies not only disorder but also an impending new order, a coming together like the confluence of two streams into a larger flow.

Consequences such as introducing new learning modalities, and forced non-attachment and engagement operate at the behavioral level. Tactics such modeling introduce both new behaviors but also compel a non-cerebral learning style. Torment/frustration demands a letting go of old behaviors and preferences. While less explicitly emphasized in Buddhist circles than non-attachment, is new engagement. The disciple in both schools is not only being asked to let go but also to show up in the here/now. In the face of confrontation, indeterminate speech, paradoxical demand, "one has to say something," Katagiri Roshi reminds us. Clearly the tactics of instruction are multi-functional, engendering both inner states and behaviors that ideally move the disciple toward awakening.

General Characteristics

The designation of distinct tactics is for analytic purposes only. In practice, these tactics rarely occur in isolation, rather they are combined. The identification of tactics allows a sketch of their shared characteristics. First, the teaching is person centered rather than scripture centered. The teaching is person centered in two ways: the audience or target of the teaching and the one who delivers it, the master. Anecdotes, stories, and personal testimony far outweigh references to scripture and commentary. The texts of Ch'an are the records of oral teachings and biographical anecdotes. The world is left with the teaching incidents of Bodhidharma, Baijan, Huang Po, Mazu, and Linchi, not commentaries on sutras. The teachings of the Forest Tradition are, at least in practice, the dhamma talks and biographies of its ajahns. The classical sources remain, but in the background. What has communicated the Forest Tradition to the public are their published works, none of which to knowledge are formal commentaries on text. By enlarge, they are discourses on practice related themes and personal tales. Biography and autobiography are prevalent. Ajahn Mun, Maha Bua, Chah, Thate, Khao, and Lee have all been subjects of such literary treatment. Scripture is cited as an additional support or to introduce a talking point. The body of the teaching is a more personal message arising out of the teacher's practice/realization. Chapter 6 is an in-depth examination of the hagiographies and biographies of the mountain and forest masters.

Second, the training in both schools is situational, pertinent to a particular moment and actor, rather than generalized truths. Ajahn Sumedho recounts his teacher's angry disapproval of his prepared dhamma talk. "Don't ever do that again." "Genuine dhamma talks don't come from a script," Ajahn Pramote tells us (Pramote 2013, p. 83). They are of the moment. Ajahn Chah himself reports Ajahn Mun's individualized instruction. Overwhelmed by the weight of disciplinary rules he found in the classic texts, Chah sought Mun's teachings, which apparently synthesized complex material into a workable practice. The teaching was both pith, but also intensely relevant to the young monk (Jayasaro 2008, p. 13). The Aside from Dhamma talks the individual interactions are also meant for the specific student. This is what the Ch'an school calls "living words" as opposed to dead discourse.

Zen interviews are private. This is misconstrued to be a necessity to hide koan answers, but such an interpretation misses the more general principle that the entire exchange is unique to that disciple and that moment. What is true today may not be true tomorrow. What is true for one student may not apply to another. In a Zen retreat, sesshin (Jp.), the author attended the teacher demanded that several students who had passed koans do them again. The moment is always new. The forest monk's bewilderment at his teacher's apparently contradictory advice to different disciples is illustrative of the situational quality of the teaching tactics. Abstract thinking is undermined in favor of an intelligence, *panna* (P.) and *prajna* (Skt.) that creatively responds to the present situation.

A third trait is immediacy. The immediate situation not only calls forth the teaching, but the teaching turns the receiver's attention to the present. The value of present-awareness is an implicit learning embedded in many of the teaching tactics. Utilization it has already been noted performs this function consistently. Many of the teaching tactics have this corollary consequence. Confrontation can shock one into the moment. Modeling of behavior requires attention to the here and now. The Dhamma is right here, right now. Method and goal are intertwined.

Immediacy as a function of teaching method also has textual antecedents. The relationship between the phenomenal word and ultimate truth is a central and complex issue in Buddhist thought. The Ch'an school can call on centuries of Mahayana philosophy that both distinguish the phenomenal from the ultimate and assert a radical intimacy between the two. Ch'an took this seriously as a working point for practice and realization. The immediate was not a distraction or defilement. Rather it was the very place of enlightenment. "This very place is the pure Lotus Land," declares Hakuin, the great Rinzai reformer (Suzuki 1994, p. 151). "We are awakened by the myriad things," Dogen asserts (Dogen 2011, p. 24). Ch'an/Zen has significant doctrinal scaffolding to support teaching tactics that drive the disciple into the immediate.

In Theravada tradition, nature and Dhamma have always been closely associated. Buddhadasa, the internationally renowned Thai monk and a contemporary of the Forest Tradition, unequivocally identifies Dhamma as the laws of nature. If you know nature, you know the Dhamma (Buddhadasa 2006). As previously quoted, Ajahn Lee became convinced that he could dispense with scriptural study in favor of the original text, nature. In conversation, a foreign monk in the Forest Tradition spoke of the immense insights he had by living alone in the forest for a year. In the nearby pond, he had received a discourse on the life cycle that no printed word could convey. If Dhamma is in nature, then directing the student to nature and the phenomenal world is an introduction to the supreme Buddhist education.

Fourth, as has been already noted, the style is pragmatic. Does the tactic work to liberate the mind? Consequences and not conformity to an idealized role of the monk or the complexities of Vinaya rules is what guides the teacher. The entire project has been labeled pragmatic-soteriological. Carrithers (1983, pp. 48–51) in his study of Sri Lankan forest monks refers to the psychological pragmatism of their practice. It aims for the ultimate but proceeds by practical consequences. This it should be noted reverses the de facto pattern of much of religious life. Attainment of the ultimate is relinquished in favor of holding everyday activity to mainstream morality is typical. Again and again the guiding consideration of both Ch'an/Zen and Forest teaching gambits is shown to be their consequences for attaining the ultimate, awakening.

In a self-revealing conversation, Ajahn Chah explains his method. First, he lets the monk do as he wishes. Second, he observes the monk's proclivities. Third, he begins to frustrate the displayed tendencies (Breiter 2004, loc. 1105). The method proceeds from permissiveness to torment. The teaching style demonstrates not only keen pragmatic considerations, but also the person-centered and situational nature of Chah's training. We hear nothing of explicit Buddhist values. The training is more boot camp than Sunday school. The value is in the outcome.

Lastly, there is an antinomian scent that wafts through the teaching style. Certainly, these movements do not violate the moral standards of society. Rather they see themselves, especially the Thai tradition, as maintaining an ascetic ideal. Nor do they launch any challenge to the socio-political order. If anything, they are forced to defend themselves against accusations of collusion with the modern state. Yet within the domain of teacher student relations, they flout the rules of social interaction and the use of language. The idealized role of the monk is overturned. There is a sense that within this field of face to face micro-interaction, anything goes as long as it passes the pragmatic-soteriological test. If pulling face, talking gibberish, absurd commands, puzzling stories and statements work to free the student from prison of delusion, all is fine because all is fair in the quest for awakening. A story circulating in Zen circles indicates that once this new role as the religious rebel has been established that too has to be over thrown. A Western disciple chides his Japanese master for bowing to Buddha images instead of burning them for warmth as in the old stories. With a smile, the old teacher responds, "I prefer to bow."

There is a palpable sense that once one has stepped out of an audience politely listening to a public talk into a more direct relationship and that one is on dangerous ground. After giving a public talk in NYC, master Sheng Yen made his way slowly through the hall. Suddenly, he turned sharply and came down the aisle toward an African American decked in dread locks standing next to me and obvious in the crowd of conservatively dressed Asian and white middle class attendees. Standing inches from the man's face, his smile suggested kindness, his eyes hawk-like. "Interested?" The fellow nodded. "Good." Compressed into the moment was both an overt friendliness but also a preview of penetrating awareness.

Summary

The teaching tactics and general characteristics of the training style of the Ch'an and Forest Tradition have been examined and shown to have marked similarities despite their affiliation with different branches of Buddhism and their temporal/geographic distance. The teaching relationship was identified as charismatic and the project as pragmatic-soteriological, meaning an overriding commitment to liberation as the core value. The tactics were identified as: utilization, indeterminate and indirect language, paradoxical speech and commands, contradiction/deconstruction, confrontation, compassionate frustration, and modeling behavior. The list is not meant to be exhaustive or exclusive but to be a categorization of what might otherwise be seen as idiosyncratic even random behavior.

Several generalizations emerge. The tactics are shared by the Thai Forest Tradition and Ch'an/Zen. While acknowledging that the teaching maneuvers are nuanced differently (different teachers, different schools), an overview reveals striking similarities. Grouped in a list of descriptive categories, the tactics have overlapping functions which drive forward the pragmatic-soteriological project. At the level of internal states, they engender shock and confusion in the actor, which is a disruption of old cognitive templates, and under appropriate conditions, a preparation for new learning. At the level of behaviors, they both teach the new and frustrate the old. Along with the common Buddhist injunction to let go, it was pointed out the tactics equally require the actor's active engagement. Both letting go and intense engagement are part of a rhythm moving toward liberation. The tactics also share characteristics that give a distinct tone to the teaching relationship of the two schools. The teaching style is person centered, situational, immediate, and pragmatic. Taken together, we hear an antinomian strain, which while never the dominant melody of two movements, both of whom remain within the Buddhist establishment, it does offer a recurrent hint of radical means to an ultimate end.

An additional note needs to be added. In dissecting the teaching tactics and their functions and labeling the relationship as charismatic, simple human truths can be lost. What seem too often to underpin the granting of charismatic authority and the willingness to endure tough teaching methods was a far more ordinary quality. What attracted Thanissaro Bhikkhu to his teacher? He was the first authentically happy man he had met. For Guoyuan, it was Sheng Yen's sense of humor. For Amaro Bhikkhu, it was Chah's spontaneity. Perhaps the basis of complex charismatic connections is the simplest of human attributes.

References

Breiter, P. (2004). *Venerable father*. New York: Paraview.

Buddhadasa, B. (2006). *Truth of nature*. New York: UNESCO.

Burlingame, E. (1999). *Buddhist parables*. Delhi: Motilal Banarsidass.

Carrithers, M. (1983). *The forest monks of Sri Lanka*. Delhi: Oxford University Press.

Chah, A. (2002). *Food for the heart*. Boston, MA: Wisdom.

Chah, A. (2005). *Everything arises, everything falls away*. Boston, MA: Shambhala.

Chah, A. (2013). *Still flowing water*. Thailand, The Sangha.

Daehaeng, S. (2007). *No river to cross*. Sommerville, MA: Wisdom.

Dogen, E. (2011). *Dogen's Genjokoan*. Berkeley, CA: Counter Point Press.

Dumoulin, H. (2005a). *Zen Buddhism: A history* (Vol. 1). Bloomington, IN: World Wisdom.

Dumoulin, H. (2005b). *Zen Buddhism: A history* (Vol. 2). Bloomington, IN: World Wisdom.

Erickson, M. (1996). *The wisdom of Milton Erickson* (Vol. 2). Bethel, CT.

Ferguson, A. (2000). *Zen's Chinese heritage*. Boston, MA: Wisdom.

Fuang Jotiko, A. (1999). *Awareness itself.* Valley Center, CA: Metta Forest Monastery.

Gou Gu. (2012). *Silent illumination*. Retrieved March 4, 2015, from http://www.tallahasseechan.com/talks.html

Hoover, T. (1977). *Zen culture*. New York: Random House.
Jayasaro, A. (2008). *Ajahn Chah's biography by Ajahn Jayasaro by the noble path*. Retrieved January 8, 2015, from https://www.youtube.com/watch#1-43
Katagiri, D. (2000). *You have to say something!* Boston, MA: Shambhala.
Kwong, J. (2007). *Breath sweeps mind*. Boulder, CO: Sounds True Audio.
Lee, D. (1991). *The autobiography of Phra Ajaan Lee*. Valley Center, CA: Metta Forest Monastery.
Leighton, T. D. (2000). *Cultivating the empty field*. Boston: Tuttle Co.
Maha Bua, B. (1995b). *The venerable Phra Acariya Mun*. Samudra Sakorn: Wat Prajayarangsi.
Maha Bua, B. (2006). *Venerable Ajaan Khao Analayo*. Udon Thani: Forest Dhamma Books.
McDaniel, J. (2011). *The lovelorn ghost and the magical monk*. New York: Columbia University Press.
McRae, J. (2003). *Seeing through Zen*. Berkeley, CA: University of CA Press.
Mills, C. W. (1940). *Situated actions and vocabularies of motives. American Sociological Review*, 5(6), 901–913.
Perls, F. (1969). *Gestalt therapy verbatim*. Moab, UT: Real People Press.
Powell, W. (1986). *The record of Tung-shan*. Honolulu, HI: University of Hawaii.
Pramote, P. (2013). *To see the truth*. Bangkok: Amarin Dhamma.
Senior Sangha. (2013). *Recollections of Ajahn Chah*. Hertfordshire, UK: Amavarati Publications.
Silaratano, B. (2009). *Mae Chi Kaew*. Udon Thani: Forest Dhamma Books.
Stevens, J. (1977). *One robe, one bowl*. New York: Weatherhill.
Stryk, L. (1994). *Zen poems of China and Japan*. New York: Grove Press.
Suzuki, D. T. (1994). *The manual of Zen Buddhism*. New York: Grove Press.
Suzuki, S. (1999). *Branching streams flow in the darkness*. Berkeley, CA: University of California Press.
Taylor, J. L. (1993). *Forest monks and the nation-state*. Singapore: Institute of Southeast Asian Studies.
Thanissaro, B. (2003). *Angulimala Sutta: About Angulimala*. Retrieved April 1, 2015, from http://www.accesstoinsight.org/tipitaka/mn/mn.086.than.html
Thate, A. (1993). *The autobiography of a forest monk*. Wat Hin Mark Peng: Amarin Printing.
Tiyavanich, K. (1997). *Forest recollections*. Chiang Mai: Silkworm Books.
Waddell, N. (2000). *The unborn*. New York: North Point Press.
Wallis, G. (2007). *The basic teachings of the Buddha*. New York: Modern Library Paperback.
Wang, Y. (2003). *The linguistic strategies in Daoist Zhuangzi and Ch'an Buddhism*. London: Routledge Curzon.
Weber, M. (1946). *Essays in sociology*. New York: Oxford University Press.
Wu, J. (2008). *Enlightenment in dispute*. Oxford, UK: Oxford University Press.

CHAPTER 4

Warriors of the Way: Practice and Path

Introduction

Buddhism is a religion of path, a way to awakening, through activities that move one down the road. Practice and path in the Ch'an/Zen and Thai forest school display some not unexpected differences, given their affiliation with different denominations, Mahayana and Theravada. Yet they also share similarities in overall style and temperament and even specific practices. This chapter examines the core practices of both movements and their understanding of the path. A vital area of practice and significant convergence has already been investigated: the teacher/student relationship and its teaching tactics. Here, we will present a broader review of practices and the Ch'an/Zen and Forest Tradition's conceptualization of path.

Among the key substantive areas of comparative investigation will be: meditation, *sila* or precepts, work, alms round or pindabat, and general living/practice situation. Differences in the conception of path itself will be elucidated. Out of this investigation, several lines of divergent overarching orientations emerge. First is an individualist/collectivist preference with the Forest Tradition identified with the former and Ch'an/Zen with the latter. Second is the preference of the Forest Tradition and Ch'an/Zen to present around contrasting principles of ascetic/discipline or wisdom/iconoclasm. Lastly, beyond variations in technique and practice forms, we will look into the shared core and guiding values of walking the way, the very temperament of the practice.

A.R. Lopez, *Buddhist Revivalist Movements*,
DOI 10.1057/978-1-137-54086-7_4

Meditation

The cardinal practice in both schools is meditation, in Pali *citta bhavana*, in Sanskrit *dhyana*. A review of meditation in the Zen and Thai tradition identifies differences in technique and application and yet also undeniable similarities. What is evident is the centrality of the practice for both the forest and mountain schools. Ch'an, of course, is the meditation sect. Ch'an's very label is derived from the Sanskrit, *dhyana,* or in Chinese, *channa*. Zen, of course, is likewise the abbreviation of the Japanese term for meditation, *zenna*. This definition and etymology plays out in the Korean version, *Soen*, and the Vietnamese, *Thein*, as well. The mountain school is, by self-definition, the meditation sect.

Likewise, the Forest movement defines itself and separates from other Thai Theravada monks on two accounts: the strict adherence to sila precepts, and adjacent ascetic practices, and the rigorous practice of meditation. As sila was the shield of purity, so meditation sword of wisdom. We must remember that both movements are self-branding and that there is a tendency to cast aspersion on their competitors as slackers in the field of meditation. Nevertheless, the commitment to meditation as the royal road to awakening is a primary and shared focus. Monks in both traditions are expected to dedicate considerable time to the practice which is in turn monitored by their teachers for progress.

Meditation in the Forest

Thai Theravada meditation can claim seniority over Ch'an. In theory, it finds its antecedents in the Pali Canon and the forests of north India and Sri Lanka. Yet the primary meditation practice of the Forest Movement, the repetition of the mantra Bud-dho, does not directly appear in the suttas. Apparently, the journey from primitive Buddhism to the jungles of Thailand is a long and sparsely documented path. The origins of the concentration practice, favored by the Forest Tradition, are difficult to determine. Records for the practices of Theravada monks, and especially forest monks, prior to the twentieth century are scant. Laundry lists of practices generally give prominence to concentration on mantras and *nimitas*, images arising in meditation, over the mention of vipassana or insight. For example, this seems to be the case in the training of Kruba Srivichai and Kruba Kao Pii two pre-modern twentieth century monks of Lanna, Northern Thailand (Sriwichai 2005, p. 5). Nevertheless, the assumption, particularly prevalent in the West, that Theravada practice is synonymous with insight meditation needs to be adjusted especially in attempting to understand the Thai Forest tradition.

Thai forest meditation generally begins with *Buddho*. The two syllables, Bud and dho, are coordinated with the in and out breath, respectively. The student is directed to maintain concentration on the coordinated breath/sound to the exclusion of all other thought and sensory input. Samadhi, concentration, lays the necessary foundation for wisdom, panna. It is argued that without the penetrative power of samadhi, insight can only be shallow. The practice of Bud-dho is a near universal among forest monks. While other methods are suggested, the recitation of Bud-dho is the favored object of sitting meditation. Ajahn Chah for instance recommends that at the commencement of a meditation period a body scan is helpful; nonetheless, he continues to advocate Bud-dho as the primary practice. This is evident in his instructions to a dying lay follower as he urges her to let go of all concerns and just be with Bud-dho (2002, p. 323).

The work *Meditations on the Word Buddho for Beginners* (2015) by Chaiwutimongol is one of the few in print discussions of this primary practice. A progressive set of levels of Budh-dho are prescribed beginning with coordination with the breath. The mantra is eventually located in the heart or solar plexus region. The systematic presentation by Chaiwutimongol fills in a gap in our knowledge, but it may be misleading. The researcher never encountered such an orderly or thorough practice of Bud-dho. This presentation may be an ideal composite. Nevertheless, in the field, Bud-dho was practiced to establish basic samadhi.

Once the mind has become unified with Bud-dho/breath, the practice becomes more subtle and even divergent. The attainment of a stable absorption in Bud-dho is a critical junction and signals a higher level of practice. Evidence for this shift is available in both recorded public teachings and in private personal instruction. In practicing at Luang Por Viriyan's wat and on his forest walks, the author reported to his meditation instructor the disappearance of both breath and Bud-dho. The instructor became hyper-alert and questioned me closely. What he heard convinced him to pause the interview while he sought further input from a superior. Ajahn Chah discusses a similar state with his dying student. When there is no Bud-dho and no breath, there is just awareness, pure wakefulness. This Chah declares is meeting the real Buddha. While the author makes no such claim for himself, it is clear that this is a critical juncture on the meditative path. Arriving at this place occasioned a fierce dispute between Ajahn Thate and his master over how to proceed (Thate 1993, p. 148). Does one turn back to further investigate the conditioned or does one leap into the unconditioned? Most, but not all, Forest masters counsel the former, while Zen urges the later. The preference for investigating the conditioned vs. jumping into the unconditioned is a significant difference between the two schools.

While the antecedents of the Bud-dho practice are most likely derived from Theravada missionary monks from Sri Lanka and the Forest monks of Southeast Asia, contemporary Forest practice also offers an innovative technique. This is significant in that Asian schools and especially the Theravada have strong conservative identity. The replication of the original confers legitimacy. An "originalist" bias runs through the Thai Forest School, so it is notable that a new meditation technique has been invented and acknowledged. Ajahn Sumedho is the most senior Western disciple of Ajahn Chah and has independently formulated a mediation based on sound, what he dubs the sound of silence (Sumedho 2007, pp. 110–117). Upon taking up residence in England far from the cacophony of Thailand, he discovered an inner buzz that could be used as a meditative focus both in and out of formal practice sessions. The willingness to innovate, albeit by a Western monk in a Western setting, suggests a commonality with Ch'an/Zen in its inventive meditation forms. Subsequently, one of Sumedho's students pointed out that this is a method expounded in the Ch'an favorite *Surangama Sutra*. The author was introduced to this method by a Chinese teacher while on retreat.

A story comes to us by way of Ajahn Pasanno about Sumedho's first meditation discussion with Ajahn Chah. Dissatisfied with a Burmese technique he had been practicing, he read of a Ch'an method taught by Hsu Yun, China's outstanding twentieth-century teacher. Chah asks if it works for him. Yes. Well that seems OK. Two vital pieces of information emerge. First, this is probably the earliest direct Ch'an/Forest Tradition contact. Second, this is Chah's pragmatic, non-sectarian approach (Pasanno 2012, pp. 8–9). The irony is that the forest Tradition's less formal approach to meditation as opposed to Ch'an/Zen actually opens the Thai school to incorporating Mahayana/Ch'an methods.

Ch'an/Zen Meditation

The first Patriarch of Ch'an, Bodhidharma, heralds the primacy of meditation in his school with his iconic gesture of sitting facing the wall at a cave above Shaolin Temple. The Soto sect to this day meditates facing the wall in a gesture of that is as symbolic of the school's lineage and spirit of practice as it is functional. A thorough review of Ch'an/Zen meditation would require volumes. However, an attempt will be made to sketch its salient features, especially those that offer comparison and contrast with the Thai Forest School. Ch'an largely lives up to its title by placing and largely keeping meditation at the heart of practice. What is equally, if not more, significant than the practice itself is Ch'an's self-branding as the Buddhism of meditation. A balanced history of the place in Ch'an in Chinese Buddhism recognizes that Ch'an by no means had a monopoly on meditation. Partisans may imply that the other monks were only bookworms or funeral monks, but in fact other schools not only practiced but produced significant meditation manuals. *The Great Stopping and Seeing*, by T'ien-t'ai Master Chih-i, is a Chinese classic on the relationship between the ever debated relationship between concentration, samadhi, and insight, *vispasyana* (Skt.) (Cleary 1997a, b, c). A meditative path is laid out that re-works instruction found in traditional Pali texts and the Chinese near equivalent texts, the Agamas. Such works anticipate the soon to emerge Ch'an school and its revolutionary re-visioning of the relationship between wisdom and concentration, prajna and samadhi.

Ch'an/Zen meditation has undergone a complex development at times as occluded as the mist covered peaks of Chinese landscape paintings. The earliest practice is *pi-kuan (Ch.)*, wall gazing, associated with Bodhidharma. The Fifth Ancestor, Hongren, counseled that after establishing awareness of the continuous flow of impermanence to blend into it repeatedly (Cleary 1995, p. 13) again we see the distinctive quality of Zen. The meditative theme of impermanence, so emphasized in Theravada and the Forest tradition, is present in Zen/Ch'an but only as a stepping stone for a direct leap into the Ultimate. Throughout Ch'an/Zen history, masters leave instructions to "look into the mind ground," "meditate on the nature of mind," and similar phrasings that suggest a turning of awareness on the mind itself. While there may be variations in the precise application of this mind on mind meditation, they fall within the general scope of Bodhidharma's direct pointing to the nature of Mind.

The Zen practice of koans, using the classic stories of Zen encounters, as a meditation theme would seem to be the sharpest difference with the Thai Forest practice. The koan meditation or kanna Zen style is unquestionably a distinguishing mark that sets Zen apart from all Buddhist schools. However, a deeper investigation reveals a surprising commonality between the initial koan exercise and the Bud-dho practice.

Skipping to the contemporary Zen practice of the Linchi/Rinzai school, we encounter a concentration meditation similar to the Thai Forest meditation. Along with the sound of one hand clapping koan, especially popular in the Hakuin lineage, is Joshu's *Mu* from his famous response to "Does a dog have Buddha nature?" This is the initial practice of Sanbo Kyodan influenced centers as well as Sheng Yen's Ch'an centers where the Chinese pronunciation *Wu* is used. The single syllable is repeated silently until in a state of total absorption the student breaks through the barrier of the dualistic mind. This mimics the Forest practice of "Buddho." Both are concentration practices that engender samadhi. At more advanced levels, the two schools part ways with the Thai school's rigorous de-construction of the body and Zen proceeding with more nuanced koan work, but the preliminary practices are nearly identical. Both Maha Bua and Hakuin Zenji agree that a penetrating awareness established through the repletion of a breath coordinated syllable is a necessary first stage of practice.

Before moving on to a review of how these techniques are enacted, there is a meditation associated with the Soto/Caodong school which finds no clear equivalent in the Thai forest tradition: *shikantaza* (Jp.), *mozhao* (Ch.). "Just sitting," the most popular translation of this approach is, according to its main promoter, Dogen Zenji, not a meditation at all, but a verification/actualization of enlightenment moment by moment. Buddha Mind is manifested again and again. Obviously, this rests on a doctrinal position of original enlightenment, and finds expression in a sitting practice that is unmatched by any in the Theravada school. However, as will be shown, within the Forest School, there are indications of an opening to this position.

Walking Meditation

After sitting meditation, walking meditation is the most valued and common meditation exercise for both the mountain and forest dwellers. Walking is among the four postures the Buddha speaks of as naturally occurring activities in which attentiveness to the moment, sati, should be practiced. Walking meditation is seen as supportive of seated practice and as a bridge to everyday life. The mind is trained in stillness and attentiveness, permitting the sustaining of the meditative state during the course of daily activities. In both Ch'an/Zen and the Forest Movement, walking meditation has been given a characteristic form.

Within the Theravada there a variety of walking styles especially among the different Burmese insight meditation schools. The Thai Forest practitioners opt for a relatively natural gait when compared with the mechanical like precision found at many vipassana centers. The pace is slow with the left arm grasping the right arm which is held down in front of the right side of the torso, eyes are open, and awareness is on the movement of the feet. Perhaps the most distinctive observable aspect is that it is performed in a single marked off track in which the practitioner paces back and forth, turning smartly at either end to re-trace his/her path.

Walking meditation in the Forest School, unlike in the Zen tradition, is a primary practice credited with bringing aspirants to awaking. Ajahn Mun, for instance, was reported to have entered arahantship while engaged in mindfully walking. A temple in Northern Thailand has preserved a walking path, which is claimed as his literal path to liberation. Ajahn Liem, the author of *Walking with Awareness: A Guide to Walking Meditation* (n.d), is especially remembered for an intense walking discipline of 10 to 18 hours a day of practice. Interestingly, there is no equivalent work in the Ch'an/Zen tradition. Forest teachers take walking seriously.

The belief in the efficacy of walking mindfulness separates the forest and the mountain. All Ch'an/Zen centers and retreat include a walking practice, but in the researcher's experience, it was never presented as a practice that could supersede seated meditation or could through its intense exercise bring about awakening. Certainly *kinhin* (Jp.) was valued beyond just a much needed leg stretch. Concentration was taken off the cushion and sustained in activity, but seated meditation remained the king. The author did hear of a head monk at a Western center who experienced *kensho,* awakening, while walking. True or not it is an anomaly in the Zen world where seated zazen has an iconic status.

The solitary monk pacing in serene silence on a path cut out from thick surrounding foliage is the image of the forest monk. Even as he may seem remote from his environment, it encloses him. The Soto Zen monk is in line with his fellows moving in a synchronized almost excruciatingly slow rate resembling a single, multi-legged insect making its way around the meditation hall. Rinzai/Linchi offers a quicker version of the practice and some Chinese temples increasing velocity in an exercise called circle running which when practiced on outdoor trails tests not only speed but nimbleness. Yet true to the Ch'an/Zen temperament, walking practice is a coordinated group activity. Forest temples will have group walking, but this was only encountered in retreats hosting numerous outsiders. One has the impression that once the visitors have departed, the monks will return to their preferred unaccompanied, studied stroll.

Meditation and Context: A Comparative View

Ch'an/Zen and the Thai Forest School each have their unique meditation techniques. The Forest School has its Buddho and Ch'an its koans. Sino-Buddhism has incorporated devotional meditations on Amida and Kuan Yin form Pure land Buddhism. *Ashuba* or the meditation on the unattractiveness of the body is a classic Theravada exercise. In an urban adaptation of a forest practice, one group would visit the autopsy room and morgue of the large government hospital in Chiang Mai for their contemplations. A senior monk of Ajahn Chah unequivocally asserted that he had never seen advanced attainment without analytical body meditation (Peeak n.d). Of equal importance, but far more attractive to Westerners, is the cultivation of metta. Loving-kindness meditation has been taken up by non-Buddhist groups and secular programs. These meditations are not unique to the Forest Tradition but are common to the Theravada, but absent in contemporary Ch'an/Zen.

In observing the actual enactment of meditation, there are obvious distinctions in style and tone of the practice. In Japanese style, Zen meditation is approached with elaborate ritualized bowing and a prescribed route through the meditation hall, *zendo*, to one's assigned seat. Meditation places are designated and laid out in a straight line, military-like pattern, executed on a small cushion or *zafu* in a precise posture. Full lotus and half lotus are preferred with knees on the tatami mat, back erect, chin in, abdomen is relax and thrust slightly forward, breathing is deep and slow, and eyes open. A meditation monitor stalks the hall correcting posture and applying the *kyosaku* stick to engender determination. Dark colored robes, worn even by laypersons, add to an almost grimly disciplined atmosphere. The highly choreographed scene denotes that zazen is, among other functions, a ritual enactment of enlightenment (Leighton 2008, pp. 167–183).

The forest practice offers evident contrast. There are no assigned seats. Posture is more casual, by Zen standards, possibly slovenly. Ajahn Chah does recommend three prostrations at the commencement of sitting. Cushions are not used. Besides attesting to the enviable lability of Thai hip flexors, the absence of support suggests continuity between everyday sitting and meditation. The looser approach to posture combines with the absence of any focus on the lower abdomen. In Zen, the *tantien* (Ch.) *dantien* (Jp.) is understood as the vital spiritual center where heaven and earth meet. Disciples are urged to "throw strength into the center." Other than as a possible option for mindfulness exercise, this region of the body receives no special attention in the Forest Tradition. Here, we have two significant differences that play out across many practices and speak to underlying attitudes. Zen is more physical and embodied. Paradoxically, it is also more symbolic or ritualized.

The Thai Forest Tradition is more instrumental but also more contextualized. The very Pali term for meditation, *citta bhavana*, the cultivation of the mind, suggests a wider project. Sitting meditation can understood as one aspect of mind cultivation. During a solo retreat at a Thai Forest wat, the Abbott suggested that too much meditation can be counter-productive. Best to mix meditation with other activities, he advised. Ajahn Chah suggests not setting a time on meditation sessions, unlike the closely regulated Zen meditation periods. Both Chah and Maha Bua propose investigating hindrances that arise in seated meditation while not engaged in formal practice. Maha Bua even counsels that wisdom develops Samadhi, a sequence that reverses the usual developmental order (Maha Bua 2005a, b, p. 13). Their perspective sees meditation as a mode of cultivation that has continuity with other practices, certainly not a ritual set aside.

An example of this attitude is the instructions given by forest monk to a group of beginners. Meditation was likened to a scientific investigation. Conditions for the experiment were to be established: silence, erect but still body, and closed eyes. The play of key variables could now be observed. The entire approach was of a sensible, pragmatic endeavor. "Let's sit down and get to know the mind," seemed to be the subtext. Instructions on posture and hand positions (*mudras*) were bare bones functional. This is not meant to imply a lax attitude to meditation practice. Forest masters are known for their strenuous day and night efforts. Rather the practice is held differently, more loosely and less compartmentalized than Zen's set piece zazen.

Ajahn Chah summarizes attitude when he corrects a student. "Actually the practice is a matter of the mind, a matter of your awareness." He frees meditation from any formal activity. "It's not the sort of thing where you have to do this, do that, or jump around a lot. It's simply a matter of your awareness" (2013, p. 77). If you are breathing, you can meditate.

Individualism and Collectivism

Under what conditions is meditation practiced? How is it organized? These questions bring us to an important difference between the two schools that is not only apparent in meditation but in many of the aspects of the life of monks from each tradition. The individualism of the forest tradition and the collectivism of Ch'an/Zen are organizing principles that distinguish the two schools. Forest practice maintains an individualized practice despite its group activities, while Zen practice is group practice organized with uncompromising uniformity with individual activities at the margin. This pattern is found in eating, sleeping, residential location, and, of course, meditation.

At Forest wats, group meditation may occur especially in gatherings for dhamma talks or chanting. However, meditation is typically a solitary event carried out in monks' private quarters or *kuti*. The schedule is looser and any time free of other obligations may be used for meditation. Meditation is largely a self-regulated activity. When doing a solitary retreat, my questions about when and how long to meditate were met with "free style" in other words up to me. By contrast, even solitary meditation in Zen centers and monasteries was conducted in the formal meditation hall or zendo. Zen teachers would occasionally remark on the value of group support for practice. Such ideas were not present or at least expressed in forest centers. Zen teachers saw a value in the synchronized group meditation (and other activities) in facilitating a dropping of self to blend with the group. This was interpreted as ego surrender.

Certainly, monastery chores and participation in group work projects were seen as ego challenging in forest wats, but this did not extend to meditation which remained a largely personal activity in comparison to Zen. There is a group pressure and reward in the Forest Tradition in regard to meditation practice. Monks are noted for more or less vigor in their endeavor of citta bhavana. However, the group effect although significant was informal. The when and how long of a meditative session was set by the individual monk or layperson on a private retreat. It was the solitary effort that was most prized again underscoring the individualism of the forest tradition.

Sila and Ascetic Practices

Sila, the moral precepts, along with accompanying ascetic practices are arguably the shaping structure for the life and practice of the monk. While it is meditation that has captured the interest of Westerners, the everyday life of the ordained is equally, if not more so, about the forms of moral and social etiquette. Indeed, there are monks who would see practicing the precepts as primary without which more elite practices such as mediation would be meaningless or impossible. For the Thai Forest School, precept and meditation are the two legs which walk the path. Guoyuan, a Ch'an monk who practiced in Thailand and at Wat Nanachat, suggested to the author that keeping the precepts was in itself a samadhi or deep concentration for the Thai monks. Abiding by the precepts is its own meditation, citta bhavana.

Sila and ascetic practices are not only different in content between the mountain and forest monks but also play different roles in self-identity. If Zen is the self-proclaimed meditation sect, then the Forest Tradition is the school of uncompromising discipline. Each school organizes its self-presentation around different poles. Forest monks embrace the ascetic/discipline principle, while the Ch'an/Zen monks the wisdom/meditation dimension. On the one hand, the forest movement teachers display wild behavior that stretches, if not the precepts themselves, then the stereotypical monk role. Teachers rail at students, utter puzzling statements, crawl like dogs, or use earthy language. On the other hand, the Forest Movement's touted claim to moral rectitude separates it from the impugned laxity of other Thai monks. On numerous occasions, the author heard Mahanikai monks, the older and larger Buddhist denomination, disparaged as failing to abide by the Buddha's rules, personally lazy and unkempt, and generally ignorant of dhamma. "Village boys in orange robes," was the characterization. Strict adherence to the Vinaya is more than a tactic of liberation. It is a marker of group identity and aligns the Forest Movement with the Theravada axiom that asceticism equals the sacred.

Along with these overarching variances in self-conception, there are differences in the forms of the disciplines worth noting. Zen precepts are based on the Bodhisattva vow and not the basic precepts of non-Mahayana schools. The ten grave violations are chanted with a brief interpretation by both Bodhidharma and Dogen included giving the classical injunctions a strong Ch'an/Zen tone. With the Meiji Restoration, monks were permitted to marry transforming them into priests. The Forest Tradition, on the other hand, follows the classic early Buddhist formulas of the Five Precepts and reverence for the Three Jewels, Buddha, Dhamma, and Sangha.

Undoubtedly, the prominent marker of the Forest Tradition's asceticism and superiority to its rivals is the *dhutanga* (thudong, Th.) or forest wandering practice. The dhutanga rules are the touchstone for the Forest Tradition's claim of an uncompromising discipline. These guidelines, formulated in the Buddha's times, are a response to a request for more austere practice than the minimal framework laid down by the Buddha. While the modern sensibility might find the monks life excessively rigorous, in

North India of the fifth-century B.CH.E., this was indeed a modest path, the Middle Way. The Buddha was faced with revolts from the "right" demanding a practice that replicated the forest dwelling ascetics encountered in his youth. The dhutanga precepts were his answer.

The dhutanga percepts are voluntary and indicate the ardent aspiration of the monk. They consist of 13 in all and are to be found in the Abhidhamma commentary, the Vimuttimagga, the Path of Freedom. The term is comprised of *dhuta*, to abandon, and *anga*, state of mind together constituting renunciation. The 13 consist of:

1. Refuse-rag-wearer's Practice (pamsukulik'anga)—wearing robes made up from discarded or soiled cloth and not accepting and wearing ready-made robes offered by householders.
2. Triple-robe-wearer's Practice (tecivarik'anga)—having and wearing only three robes and not having additional allowable robes.
3. Alms-food-eater's Practice (pindapatik'anga)—eating only food collected on pindapata or the almsround while not accepting food in the vihara or offered by invitation in a layman's house.
4. House-to-house-seeker's Practice (sapadanik'anga)—not omitting any house while going for alms; not choosing only to go to rich households or those selected for some other reason as relations, and so on.
5. One-sessioner's practice (ekasanik'anga)—eating one meal a day and refusing other food offered before midday. (Those Gone Forth may not, unless ill, partake of food from midday until dawn the next day.)
6. Bowl-food-eater's Practice (pattapindik'anga)—eating food from his bowl in which it is mixed together rather than from plates and dishes.
7. Later-food-refuser's Practice (khalu-paccha-bhattik'anga)—not taking any more food after one has shown that one is satisfied, even though lay-people wish to offer more.
8. Forest-dweller's Practice (Araññik'anga)—not dwelling in a town or village but living secluded, away from all kinds of distractions.
9. Tree-root-dweller's Practice (rukkhamulik'anga)—living under a tree without the shelter of a roof.
10. Open-air-dweller's Practice (abbhokasik'anga)—refusing a roof and a tree-root, the practice may be undertaken sheltered by a tent of robes.

11. Charnel-ground-dweller's Practice (susanik'anga)—living in or nearby a charnel-field, graveyard, or cremation ground (In ancient India, there would have been abandoned and unburied corpses as well as some partially cremated corpses in such places.)
12. Any-bed-user's Practice (yatha-santhatik'anga)—being satisfied with any dwelling allotted as a sleeping place.
13. Sitter's Practice (nesajjik'anga)—living in the three postures of walking, standing, and sitting and never lying down (Khantipalo 1994).

The austerities strike at all basic comforts and consoling pleasures. As Thanissaro Bhikkhu, a Western disciple of the forest monk Fuang Jotiko, points out these practices are foremost physical, but they aim for and necessarily entail psychological transformation. The practitioner learns to live with little while weakening the draw of pleasure. Aspirants are warned against ego aggrandizement by show of these sacrifices and those with hateful characters, dosa, are discouraged from taking them up (Thanissaro 1994).

Dhutanga practices were common for the first (Ajahn Mun, Ajahn Sao) and second generations (Maha Bua, Ajahn Lee) of the Forest Movement. This is less so for the more sedentary third generation. The forest walk has archetypical meaning for the Forest Tradition and its practice even if modified into a trek invokes the signature practice of the forest monk. Amaro Bhikkhu has published a journal of one of his walks (Amaro 2011). The researcher accompanied Luang Por Viriyang, a disciple from Mun's inner circle, on a "camping" walk that he holds for foreign students who have received meditation instruction from his lineage. Participation in this neo-thudong serves as an initiation into the world of the forest monk after the disappearance of the forest.

The everyday practices of the Forest monk living in relatively civilized temples incorporate a number of the dhutanga practices, such as those concerning food, even as the sedentary life excludes other dhutanga practices. Although a majority of forest monks do not adopt all 13 austerities, the 13 practices function as standard, the ideal template, of the good monk. Aside from advanced meditative practice, commitment to dhutanga austerities is the most referenced indicator of the monk's spiritual status.

Despite these distinctions in self-presentation and substantive practice, a deeper look reveals a more complex relationship with sila than one might assume. Their shared orientation toward proximate awakening through self-cultivation under the guidance of the charismatic master exerts a tendency toward a less formulaic adherence to sila. A case in point is a pith teaching offered by Ajahn Mun. The young Ajahn Chah approached Ajahn Mun with his turmoil over how to keep the complex and extensive rules found in the texts. A pragmatic simplification ultimately relying on the heart is counseled. Ajahn Jayasaro (2008, p. 13) explains Mun's advice on based on two principles: intelligent shame and fear. While sila is never discounted, it is wisdom applied skillfully that makes precept a workable training. On another occasion, Ajahn Mun acknowledges that he does not keep the 227 monk precepts but only one precept. What is it? "The mind," Mun answered (Maha Bua 1995a, p. 221). One could imagine a similar reply from Bodhidharma.

The Forest Movement, therefore, has a complex stance on sila. There is a tension between the Thammayut adherence to precept and the charismatic/practice axis of the forest monk. The founding of the Thammayut sect by King Mongkut was occasioned by his dismay at the moral laxity of the Sangha of his time. The very establishing raison d'etre of the Thammayut was in significant degree a revival of the moral rectitude and ascetic ideal counseled in the Vinaya. Not only in its formative years, but even today the keeping of the precepts is invoked by Thammayut monks as their identifying trait. Strict adherence to the Vinaya is more than a tactic of liberation. It is a marker of group identity. Yet the insistence on the living experience of the Dhamma and its personalization pull for an approach that echoes Zen.

It should be noted that not all forest monks are Thammayut nor are all non-Thammayut monks lax in discipline. For instance, Ajahn Chah ordained as Mahanikai monk and remained in the order, unlike others who re-ordained when they became students of Ajahn Mun. Yet Chah maintained a strict observance of the precepts and according to some influenced a tightening of discipline within the Mahanikai.

Aside from distinguishing the forest monk from scholar/urban monks and local village monks, the prominence and rectitude of the precepts are a major contrast with Ch'an/Zen. This does not imply that Ch'an/Zen temples are freewheeling affairs, but the emphasis on keeping the precepts is not foreground. Ch'an's identity is not as organized around sila. The keeping of strict precept was admired by Ch'an informants, but Ch'an tends to place less emphasis on sila. Even as precepts such as celibacy are maintained the Chinese monks they was not granted enhanced respect by the laity on the basis of precepts. Perhaps the exceptions to this were the practice of rigorous retreats and the Ch'an pilgrimages of walking/prostrations in which steps alternate with full body prostrations. On the streets of Korea, monks who had survived killer monk retreats were reverentially touched by laypersons. However, this can be seen as respect for arduous practice rather than adherence to Vinaya-based precepts. They were heroic efforts not a pure way of life.

Everyday Life, Temple Life

If the dhutanga practice is the badge of the Forest School, it is not the only practice that separates from Ch'an/Zen. Several areas of observable activity mark off the forest monks from their brethren. These were both visible to the villagers and were pointed to by the monks themselves. They comprise the basic activities of everyday life: procuring food or pindabat (alms round), consuming food, work, and residential quarters.

The monks of Northeast Asia do not go out for alms rounds, although this is not entirely the case. In response to the question do Chinese monks do pindabat?, a Ch'an monk answered, "No but yes." There are some districts of the mainland where going for alms continues, but primarily it is a ceremonial event. On special occasions, a pre-advertised going for alms is held. Zen style "pindabat" is a symbolic enactment of original Buddhism, rather than a daily intercourse with the lay community. The Buddha's presumed intention in mandating alms round and the prohibition against monks growing their own food was to foster a mutual dependency between the ordained and lay communities. Forest monks more than once reflected on the support given by the lay community on their pindabat rounds and could subsequently feel shamed by any laxity in their practice. The centrality of this practice for the Thai monk and its absence from the Ch'an/Zen world was uniformly noted by East Asian Mahayanists. Asian monk informants generally admired the practice life of the forest monks. In *Memories of a Monk*, a travel log by Ch'an monk Guoyuan reflecting on his year in Thailand, he sees Thailand as "a validation of what is possible when Buddhism is fundamental and pervasive part of society" (Gouyuan n.d). Forest monks living in South Asia could live a life closer to the Buddha's prescription, and in his view are embraced by the larger society.

The Thammayut branch, including forest monks, take only one meal a day and always before noon. This is softened by an early evening dispensation with regards to liquids in which chocolate is mercifully included. Ch'an/Zen monks eat two official meals and one unofficial dinner diplomatically designated as medicine. Finally, Chinese monasteries are strictly vegetarian. This is not unique to Ch'an but is a feature of Sino-Buddhism and perhaps speaks of the fervor of early Mahayana. On the other hand, the Buddha's proclamation to take what is offered suits the pindabat-dependent world of the forest monk. If precepts and pindabat evoked admiration perhaps even longing, then diet provoked shock. Chinese monks are strictly vegetarian and Zen in West is functionally vegetarian. A Ch'an monk recounts with almost horror when confronted with a meat containing offering. Did you eat it? "I politely put it aside."

The Zen monk lives without privacy. Sleeping and eating occur in close quarters. Especially during intense mediation retreats such as *Rohatsu sesshin* held in early December, monks will sleep, eat, and meditate on adjacent tatami mats. If not meditating, monks go to sleep and rise at the same time. The meditation hall monitor will check each participant before lights out in a routine only approximated in the West in the military or in prisons. Likewise eating is on one's mat, seated on a meditation cushion. Serving of food is ritualized and systematic, with hand signals use to regulate the amount of the serving. Conformity facilitates ego surrender and allows for total engagement in spiritual work. The Japanese adage "The nail that sticks out get hammered in" is made tangible in Zen retreats.

Chinese practice is softer. Monks sleep in dormitory rooms often with two persons to a room. Eating is in a dining hall at tables rather than sitting on meditation cushions. Servers move along the tables and in some instances monks serve themselves buffet style. The relative casualness of Ch'an temples when compared to the Japanese is illustrated by the experience of American Zen students practicing a Japanese style while visiting a Ch'an Temple in China. Without bells and bowing, people enter the meditation hall and take their seats. The abbot removes his dentures and places them on the Buddha altar. Sheng Yen was heard to joke that Japanese like to turn every mental obstacle into a physical trial. Of course, Ch'an practice is rigorous and meetings with the teacher are challenging. Yet it is fair to note that within the larger Ch'an/Zen school there are variations.

Nevertheless, even Chinese temples evidence a more collective approach than the Thai Forest monasteries. Typically, the Thai Forest monk has his own quarters where he sleeps, bathes, chants, meditates, and eats. Groups practice does exist, usually consisting of morning/evening chanting, dhamma talks, and some more formal meals where the pindabat offerings are distributed based on seniority and consumed together. The forest monk, however, always has his retreat kuti where he can step out of enmeshment in group life. The Zen monk has no comparable option.

Among the activities at times overlooked in popular imaginings of ordained life is the place of work. Yet work, primarily manual labor, is central to the monk's life. The regard for work by the Thai Forest Movement and Ch'an/Zen suggests wider attitudes about practice and everyday life, as well as, containing a key difference with wide ramifications. The Zen tradition boasts of its work ethic, and Theravada monks are thought to live a life free of work. Both images are less than fully accurate.

Alms secured by pindabat rounds and gifts of food and supplies, combined with the donated labor of villagers to build and maintain buildings, would seem to exempt forest monks from significant manual labor. Although the contributions of lay supporters are vital, it does not account for all the needs of the monk community. During its earliest and most peripatetic phase, forest monks did little work, a point for which they were criticized, but once they had settled down, they were soon pressed into labor. Ajahn Chah seemed to have had endless building projects especially road constructions linking his remote wats with the outside world. Ajahn Sumedho, recently arrived at Chah's wat, complained that he had come to meditate not break rocks. Craftily Chah agreed and allowed Sumedho to retire to his kuti as he admonished his monks not to think badly of the new American monk. Lasting only a few hours, Sumedho emerged and sheepishly joined Chah's chain gang.

Maha Bua's temple, Wat Palad, has two classifications of work. One was patibat or practice work of tasks that were standard for wat maintenance or projects designated by the abbot for temple improvement. These tasks were mandatory. Another category of work was voluntary and might include projects initiated by individual monks. Along with heavy manual labor, there were of course administrative tasks. As forest temples set roots their involvement with the locals also grew. Monks might take on planning and supervisory roles in village projects. In the Wat Nanachart areas, monks entered the local schools, to establish programs for modernizing education. Like their Zen brother and sisters, Theravada monks are encouraged to understand work as an integral part of spiritual life. A forest abbot suggested to me that work was a necessary adjunct to meditation. Our private interview was concluded not with a transfer of a robe and bowl but of a broom. "Let's sweep!" Together, we swept the wat for several hours before I returned to my kuti for more meditation. Contrary to outsider assumptions, work is a regular and valued dimension of the forest monk's life.

If Zen is distinguished by any practice beyond meditation that practice is work, especially manual labor. Academia and the reading public may know Ch'an/Zen for its poetry, but the school produces far more masons, carpenters, and cooks than poets. Zen training meant learning to skillfully labor. Suzuki Roshi founder of San Francisco Zen Center was a mason. Dogen wrote on cooking with a Zen mind. Cooks are ranked high in the monastery hierarchy. Work was expected to be performed mindfully but also with vigor. Through one's self into it was the instruction. "Work is zazen," was the slogan. Zen's commitment to work was crystallized by Master Baizhang in eighth/ninth-century China. The story is that his monks seeing his advanced age hid his tools. Unable to locate them, he refused to eat. "A day without work is a day without eating." The commandment stuck to this day. His monks relented, and Zen's insistence on work was cemented (Cleary 1997a, b, c, p. xi). This view has been contradicted by Buswell whose Korean monastic experience suggests that monks perform minimal labor (Buswell 1992, p. 220). What is undeniable is that Zen presents itself as hard working. In its self-representation, Zen is "the no work/no eat school."

The work attitudes of the forest and mountain monks complement one another with one glaring exception: the production of food. The Buddha established multiple rules, recorded in the Vinaya, forbidding the growing of food or keeping livestock, thus keeping his monks dependent on the lay population for their sustenance. Food could not be stored. Direct gifts had to be consumed. Prevented from stocking up, monks had to maintain an ongoing relationship with the laity. The possibility of being hermits or an autonomous cult was blocked. Forest monks, in my observations, conform to these 2500-year-old regulations strictly. Ch'an/Zen suspended this prohibition. The production of food is a major under taking in the mountain temples and one that affords independence. Zen lore attributes their survival of the Emperor's persecution of Buddhism to their crop growing. If Zen monks could not seek alms in public, they could still eat their mountain grown foods in private. The story glorifies and sanctifies work within the Ch'an/Zen world. One contemporary scholar of Ch'an, while watching uniformly dressed monks working the rice fields communally, suggested that they may be the only communists left in China! (Porter 2009, p. 181).

It is tempting to define schools by their doctrines and religious practices, but above all they are persons living in community. At an arranged meeting between Zen monks and Benedictine monks, they found little common ground until they began to discuss how to place shoes at the door. Everyday activities can define the religious life. For forest and mountain monks, the significant contrasts are less located in Buddhist philosophy or formal practices but in the everyday actions of temple life.

The Structure of the Way

Buddhism is a Way. Perhaps more than any religious tradition, Buddhism purports to be not primarily a cosmology but a path out of the human predicament. What is the working nature of the path in the mountains and the forests? The Indian concept of path or *marga* underwent significant Chinese reinterpretation. On the other hand, the Thai forest tradition has not been completely stagnant either in its understanding of path. Thai masters have ventured flexible reinterpretations of standard Abhidhamma/Theravada models of path. The notion of path can be viewed at two levels. First, what are its stages and its pattern of progress. Second, what is place of path within the overall Buddhist view. Again, we begin with significant differences only to find that each school leans, even if unintentionally, toward the other.

The Ch'an/Zen vision of path reflects the transformation of the Indian *marga* into the Chinese Tao or the Way. The change is not merely in the word but in the rich associations and the philosophical reorientation communicated by the term Tao. The Tao is not merely a path to get somewhere, it is the ultimate reality being sought. This union of meanings invites the conflation of path with goal. Path is no longer a set of preliminary stages, but is identical with the goal. Dogen's asserts that path and practice are not separate but constitute the moment by moment actualization of enlightenment. To separate the two is heresy. Famously, Dogen declares that we do not practice to become Buddha, we practice because we *are* Buddha. In a personal communication, Ch'an Master Sheng Yen when questioned on the point of why practice put it this way, "If you are a Buddha you are practicing." Practice is Buddha in action.

What is proposed goes beyond mere philosophical view; it goes to the essence of practice. Dogen's just sitting or *shikantaza* and his grand uncle Hongzhi's *mozhao* or silent illumination are direct openings to the nature of mind. The meditation is enlightenment. As contemporary Zen Master Deshimaru puts it, this practice is not based on mindfulness but on an immediate, holistic grasp of *ku*, emptiness, the true nature of reality (Deshimaru 1983, loc. 646). An understanding of path as the goal, practice as the goal, is largely absent in Theravada teachings. The purpose of practice is to get to the other shore. Practice and path retain their linear relationship with the other shore. In contrast, a contemporary Korean Soen (Zen/Ch'an) female master declares there is no other shore. It is all right here. Reb Anderson, a contemporary Western Zen teacher, when asked by a Tibetan lama about the stages of practice declares we have no stages. A Western Zen student expresses a wish to take his practice to a deeper level. There is no other level his teacher laughs.

Zen advertises as a one step to enlightenment path. The realities of practice, however, are different. Koan practice, kanna Zen, entails the progressive working through collections of koans. One style represented by Rinzai master Sasaki Roshi is called ladder Zen to denote the step-by-step climb through koans. Ch'an Master Sheng Yen once spoke of a stage of 10,000 peaks after an initial enlightenment experience. What is that? It is the 10,000 peaks you still need to climb. Soen Master Chinul, a towering figure in Korean Zen, coined the phrase "sudden enlightenment and gradual cultivation" to indicate that there is significant work after an enlightenment experience (Buswell 1991, p. 101). Classic Ch'an while mainly eschewing talk of stages does loosely identify three movements toward full liberation: detachment, detachment from detachment, and return to the world. Of course, there is also the famous Ox Herding Woodcuts that depict a path up the mountain and back. All Zen/Ch'an centers with which the researcher had contact recognized a de facto progressive path, despite an official assertion of sudden enlightenment. Apparently, enlightenment may be a sudden, but it is not quick.

Theravada, on the other hand, offers a far neater model of path. Theravada thought divides the path into four stages with accompanying attainments. This is gospel but is it the living practice? Ajahn Chah declares the stages to be fish sauce; at best, they give flavor to a journey. Maha Bua specifically questions certain characteristics of the sotapanna, finding that a wish for domestic life and family seems, in his teaching experience, to be compatible with the first stage of awakening (1998, p. 124). (This does not entirely contradict the Abhidhamma.) Apparently, the human process does not always fit the book and the forest masters being close to living practice recognize what escapes the scholars. In recent years, forest teachers such as Ajahn Anan and Amaro Bhikkhu have advocated a practice focused on sotapanna, while Maha Bua has taught extensively on the Arahant *marga* or path.

However, in general, Forest teachers have little say on an overview of the path. They prefer to discuss the various hindrances, mental factors, and the subtleties of practice rather than locate the disciple on the road. Stages prior to full enlightenment are less defined in practice than they are in the book. For instance, we know of Mae Chee Kaew's attainment of liberation, but up to the final phase of her work, she is depicted as plagued by significant delusion and in conflict with her teacher. We hear nothing about her prior tastes of Nibbana as predicted by Pali Buddhism. Where is the progressive stability and maturity one would expect from a practitioner on the verge of total liberation? Her dramatic story does not fit the Theravada template. Rather it is reminiscent of Zen tales of sudden awakening as a resolution of an existential crisis.

As mentioned, the one stage of enlightenment that does draw the attention of Thai ajahns is the first, sotapanna. Maha Bua's re-conceptualization of the fetters dropped at this attainment and their implications for the stream winner reflect an approach grounded in practice rather than text. The contention that householder life, specifically sexual life, is compatible with the sotapanna path obviously makes the Way more layperson friendly. Sotapanna attainment is singled out for its accessibility by at least two Thai Forest ajahns. Ajahn Anan presents the sotapanna as a viable goal for the many lay practitioners who come to his wat. Likewise Amaro Bhikkhu, a Western monk in the Ajahn Chah lineage, has suggested that as the arahant may be too lofty an objective for the average practitioner, sotapanna is still within reach. These two teachers are third-generation forest practitioners who have a significant lay following, both Thai and Western. Their highlighting of the sotapanna may reflect a view of the path, which while not abandoning the classical formulation signals the movement's evolution from an exclusively monk constituency concerned with arahantship to an accommodation with laypersons where a lesser attainment is pertinent.

Their handling of the way reflects two tendencies that arise out of a practice grounded orientation combined with their revivalist fervor. On the one hand, their revisions are pragmatic and non-dogmatic. On the other hand, they are also radical in making the path/goal intimate. The disciple is warned not to take too seriously abstract formulations of path, and yet to see path and goal as proximate. The result is to keep the student focused on the immediacy of the path while keeping the soteriological objective present. "The path is now, the goal is now" could be a Zen slogan, especially in the Caodong/Soto school. At least one forest teacher,

Ajahn Sumedho, seems to edge toward this view. In a Dhamma talk given during a long retreat entitled "The End of Suffering is Now," he brings the goal into this very moment (Sumedho 2005). How are you creating yourself in the present? This is the redirection of attention he proposes to his audience. He couples this with directing attention to the awareness that know which is always in the now. In being this, knowing capacity is the end of suffering, in being what is known is the reproduction of suffering (pp. 88–90). The goal is no longer positioned at the end of a stage-structured trail. Although this teaching is not to be found with other teachers in the tradition, it illustrates that when forest ajahns innovate, it tends to bring them closer, even if inadvertently, to their Zen brothers and sisters. This will be discussed further in Chap. 8.

The Spirit of the Way

The contrasts in the forest and mountain approach to path and practice share the spirit with which the journey is engaged. We have seen that meditative practices are equally central to the practice, that they have an emphasis on samadhi, and yet differ at the more advanced stages. Both movements integrate work into practice with a difference in the manual labor permitted. The place of the precepts in public identity also differs. These complexities are transcended by a shared spirit of militant commitment to the war on samsara and the campaign for liberation. In both the Thai Forest Tradition and Ch'an/Zen, warrior metaphor is dominant. The practice is framed as a battle against the forces of delusion. Like storming an enemy fortress, the walls of Nibbana need to be scaled, opponents vanquished, and the citadel of enlightenment secured. Disciple's both lay and ordained must have the heart of a warrior and be imbued with a fighting spirit. What allows for ultimate attainment? Ajahn Peeak tells us it is unrelenting effort, meditating longer and rising earlier (Peeak n.d, p. 173).

The warrior culture of Zen is legendary from the kungfu of Shaolin temple to the sword play of the samurai immortalized in hundreds if not thousands of action films and TV programs. The connection is not limited to pop culture but can be traced throughout the historical length of Ch'an/Zen. Whatever the veracity of the story, Bodhidharma the transmitter of Zen to China is held to be the teacher of inner energy exercises to strengthen martial arts skills of the temple's monks. The most famous swordsmen of pre-modern Japan explicitly acknowledge their debt to

Zen teaching in works such as *The Sword of No-Sword* by Tesshu (Stevens 2001). What is critical, however, is that the confluence of Zen and the martial arts follows from the warrior spirit already present in the no nonsense training of direct pointing at reality.

Martial metaphors are everywhere. Ch'an/Zen meditation halls greet the visitor with statues of Manjushri, the bodhisattva of wisdom, and his flaming sword. The stick wield by the hall's monitor is Manjushri's weapon. The public back and forth between master and students, shosan, is termed dharma combat. Contemporary teachers urge their students on like soldiers entering battle. Roshi Glassman spoke of cornering the ego like a rat and then ... Yasutani Roshi's instructions speak of doing battle with the ego and the final days of retreat demanding an all-out in the push for victory. Even just sitting is for Yasutani facing death every moment (Yasutani 2002). For Kodo Sawaki, Roshi Zen was a battle cry. Subsequently, however, he rejected militarist Zen, yet his approach to practice was uncompromising (Uchiyama 2014, p. 47). Walk the path, we are urged, not as the bodhisattva of patience and compassion but as a bodhisattva warrior.

The cultural image of the Theravada monks is of self-restraint and equanimity. Thai soap operas portray the monk as standing off to the side, impassive and silent. The Thai Forest Movement pictures its saints as combatants. Maha Bua entitles his biography of Ajahn Khao, *A True Spiritual Warrior*. He portrays Khao's journey as a military campaign. Breakthrough is framed in military language. The thudong, the forest monks' signature practice, is facing death and calls forth the warrior spirit. In Tiyavanich's *Forest Recollections* and *The Buddha in the Jungle* (1997, 2003, pp. 75–109), the fear of wild elephants, tigers, not to mention illness, faced by the forest monks is vividly referenced. Here, samadhi is not a temple bound attainment, but the capacity to sit still, in mind and body, as a tiger prowls one's encampment. For all the famous forest monks, we are left to wonder how many died alone and anonymous. This is not mere literary fancy but imbues the real life practice of the forest movement. As has been mentioned, newcomers to Ajahn Chah's wat were greeted with "Have you come here to die?" Dhamma means ego death. Chah tells us that if you are sick take medicine and lie down. If you do not improve, then just die. To live the Buddha Way is to face death without flinching.

The warrior spirit, beyond specific doctrine and practices, connects the forest monks with the Buddha. Conventional presentations of the founding father typically show a being with a perfect hairdo sitting or standing in gentle composure, but the rugged truth may be quite different. In a passage remarkably human and revealing and perhaps unique in religious literature, the Buddha tells his monks what it is like to practice alone in the jungle, where every rustle of a leaf may mean the approach of a predator (Thanissaro 1998). Likewise Zen's identification with the warrior caste of Japan replicates the Indian caste heritage of the Buddha. As one forest monk told the author, "After all, the Buddha was a warrior for 35 years before being a monk."

Summary

Path and practice for the forest and mountain school were compared revealing similarities but also differences. In both schools, meditative practice is the royal road to awakening and strongly concentrative practices are featured. The Forest School in general has a less ritualized approach to meditation and more varied exercises, and functions as an investigative training of the mind rather than a reified gesture or leap into the absolute as Zen tends to imply. Additional substantive differences in practices are the thudong and daily pindabat rounds of the forest monks and the vegetarianism and food growing of the Ch'an monks.

A critical variance, however, is the degree of identification with sila. Where the forest monks self-identify with their meticulous discipline, Ch'an/Zen tends to advertises its inscrutable wisdom. Their identities coalesce around different poles: asceticism and wisdom. The vision of the Buddha's sacrifice in the jungle contrasts with Bodhidharma's iconoclastic interview with the Emperor. An additional pattern is the individualist/collectivist difference that runs through the organization of temple life. On the collectivist/individualist contrast, the forest tradition displays a more individual style of monastic life, while Ch'an/Zen is far more routinized in its meals, sleeping, meditation practice, and teacher/student contact. The less forgiving climate of Ch'an mountains may have encouraged a collectivist response, nevertheless, each tendency is a functionally autonomous style that goes beyond the material conditions. Ch'an/Zen overlooks no opportunity to routinize temple life and practice. The forest monks proudly mimic the individualism of the early sangha.

The explicit conceptualization of the path differed with Ch'an/Zen committed, at least officially, to its sudden enlightenment model and the forest monks holding to the classical Pali/Theravada stages.

However, aside from the particulars of techniques and models of the way, there is a common connection in the spirit of the practice and the insistence on the primacy of the soteriological project. With the meditative arts as its core weapon, both Ch'an/Zen and forest monks are warriors on the road to awakening. They trumpet a martial do-or-die mentality. In this, the ultimate battle, they appear as siblings in a common struggle.

References

Buswell, R. (1991). *Tracing back the radiance.* Honolulu: University of Hawaii Press.

Buswell, R. (1992). *Zen monastic experience.*

Chah, A. (2002). *Food for the heart.* Boston, MA: Wisdom.

Chaiwutimongkol, N. (2015). *Meditation on the word Buddho for beginners.* copyright Chaiwutimongkol, Kindle edition.

Cleary, T. (1995). *Minding mind.* Boston, MA: Shambhala.

Cleary, T. (1997a). *Stopping and seeing.* Boston, MA: Shambhala.

Cleary, T. (1997b). *Teachings of Zen.* Boston, MA: Shambhala.

Cleary, T. (1997c). *The five houses of Zen.* Boston, MA: Shambhala.

Gouyuan, F. (n.d). *Memoirs of a monk's journey to Thailand.* Self-published.

Jayasaro, A. (2008). *Ajahn Chah's biography by Ajahn Jayasaro by the noble path.* Retrieved January 8, 2015, from https://www.youtube.com/watch#1-43

Khantipalo, B. (1994). *With robes and bowl.* Retrieved February 3, 2015, from http://www.accesstoinsight.org/lib/authors/khantipalo

Leighton, T. D. (2008). *Zazen as an enactment ritual.* In S. Heine & D. S. Wright (Eds.), *Zen ritual.* New York: Oxford University Press.

Liem, T. (n.d.). *Walking with awareness.* Ubon: Wat Pah Nanachat.

Maha Bua, B. (1995a). *The venerable Phra Acariya Mun.* Samudra Sakorn: Wat Prajayarangsi.

Maha Bua, B. (2005a). *Arahattamagga, Arahattaphala.* Udon Thani: Forest Dhamma Books.

Maha Bua, B. (2005b). *Wisdom develops samadhi.* Udon Thani: Forest Dhamma Books.

Peeak, A. (n.d.). What it takes to reach the goal. In *Forest path* (p. 173). Ubon: Wat Pah Nanachat.

Porter, B. (2009). *Zen baggage.* Berkeley, CA: Counterpoint Press.

Sriwichai. (2005). *A history of Kruba Sriwichai.* Chiang Mai: Sutin Press.

Sumedho, A. (2007). *The sound of silence.* Somersville, MA: Wisdom Publications.

Thanissaro, B. (1994). *Customs of the noble ones.* Retrieved April 9, 2015, from http://www.accesstoinsight.org/lib/authors/thanissaro/customs.html

Thanissaro, B. (1998). *Bhaya-bherava Sutta: Fear and terror.* Retrieved April 9, 2015, from http://www.accesstoinsight.org/tipitaka/mn/mn.004.than.html
Thate, A. (1993). *The autobiography of a forest monk.* Wat Hin Mark Peng: Amarin Printing.
Tiyavanich, K. (1997). *Forest recollections.* Chiang Mai: Silkworm Books.
Tiyavanich, K. (2003). *The Buddha in the jungle.* Chiang Mai: Silkworm.
Yasutani. (2002). Shikantaza. In D. Lori (Ed.), *Just sitting* (pp. 51–54). Somerville, MA: Wisdom.

CHAPTER 5

Awakenings

Introduction

The heart of the Thai Forest and Ch'an/Zen enterprise is awakening. Enlightenment is the goal of the soteriological project, the fount of charisma, and the sacred thread that connects the movement to the Buddha, and teacher to disciple. Religious studies have attempted comparisons of religious experience. The mystical moment with its ineffability has been the most tantalizing and yet the most elusive. Its personal immediacy and rejection of language as an adequate carrier renders the mystical marginal to a field seeking to study conformity to social norms, recurrent processes, and language-based logically coherent communication. To ignore the mystical, however, comes at a price especially when addressing traditions that organize around seeking, attaining, and verifying the mystical moment. To avoid the enlightenment event in the case of the Thai Forest Movement and Ch'an/Zen is to leave an elephant in the room. While the practitioners of the mountain and the forest have their relationships with institutions and the sweep of cultural traditions, ostensibly at least, enlightenment is the guiding star. As Amaro Bhikkhu stated it in conversation, "Nibbana is what it's about." We will explore its nature, experiential and dynamic, and its functions for each school.

A.R. Lopez, *Buddhist Revivalist Movements*,
DOI 10.1057/978-1-137-54086-7_5

Enlightenment: One or Many?

Understandings of mysticism across religions can be seen as falling on a spectrum. At one end is a socio-linguistic position that reduces the mystical to the beliefs and language of the tradition in which the event occurs. Christians have Christian experiences; Hindus have Hindu experiences, and so forth. In some forms, this view denies even the possibility of any meaningful "beyond words" knowing. Each tradition is distinct and particular. You are your lexicon. At the other extreme of the spectrum is the universal-inclusive stance that asserts that all roads lead to the same peak with the landscape only appearing varied during the ascent. Aldous Huxley's (2009) writings on the perennial philosophy convey this vision to a popular audience.

Between these endpoints are a number of qualified positions that suggest that while not all religious experiences are the same, there are varieties of mysticism that align with one another. There is in certain cases a common cross-denominational phenomenon, a resonance suggestive of an identical core. Any affinity can be inferred through two dimensions: process and content. By process is meant what is being done or practiced. By content is meant what is the language, especially metaphor, employed. The assumptions are simple. If you have a species that does something similar to its body mind and talks about the results in near identical language, we can conclude with some confidence that the experiences are a near match. This of course implies that intervening cultural variables while accounting for differences in expression are not decisive determinants. How does this apply to the Thai Forest Movement and Ch'an/Zen?

Initially, it might appear that these two movements are obvious candidates for a shared awakening experience. There practices are not radically different. While our close examination has revealed variations, they are both practice similar disciplines that would make understandable similar outcomes. Nonetheless, they represent two different branches of Buddhism, Northern and Southern, Mahayana and Theravada, with different ways of approaching and holding enlightenment. Zen, of course, touts its sudden enlightenment, while the Theravada claims the authenticity of its gradual approach. One need only listen to the sub-text of "My enlightenment is higher than yours," in Theravada/Mahayana exchanges to realize that the relationship between Thai Forest and Ch'an enlightenment requires careful consideration. Is there a Northern and Southern enlightenment?

Before narrowing our focus to the two movements under study and the actual experiences described, we need to look at the larger Buddhist heritage of both Thai Forest and Ch'an/Zen. What is enlightenment purported to be in each branch of Buddhism? Theravada offers explicit descriptions of the content of the event and its place in the process of liberation. Mahayana texts with their more grandiose and cosmological concerns tend to be less specific about what experientially constitutes the passage into Buddhahood or the upper regions of the Bodhisattva path. Theravada and Mahayana talk about enlightenment dissimilarly. The sutras and the suttas articulate differences but also intriguing convergences.

Whatever their divergent definitions, enlightenment is the signature event for both Theravada and Mahayana. Without awakening, there is no liberation. Without liberation, the soteriological project of both the forest and the mountain is canceled. Nevertheless, it is in the nature of awakening and the status it confers where we see distinctions. In the Theravada tradition, enlightenment, at least at the level of the arahant, is termed the end of suffering and the eradication of defilements. These two sides of the same coin, the end of both suffering and defilements, characterize the Theravada's descriptions of liberation in its suttas, classical commentaries, and most contemporary expositions. In enlightenment, the twin Theravada preoccupations, suffering/defilement, attain final resolution.

Theravada thought marks out a series of distinct steps to final liberation. Beginning with *sotapanna* or streamwinner, and followed by *sakadagami* or once-returner, then the non-returner or *anagami*, and finally the full enlightenment of the arahant. Theravada counts ten fetters in all with specific defilements eradicated at each stage of attainment. Therefore, the stages are progressive not only in relation to the prior stage but in their systematic removal of fetters. The dimensions used to define the process reflect the Theravada concern with purity and are quite alien to Ch'an/Zen discussions of enlightenment. Taken together, the stage model, the removal of fetters, and the confirmation of core doctrines constitute a mainstream Theravada treatment of enlightenment.

The Mahayana discussion of enlightenment does not offer the same closure or cover the same territory. While taking the awakening of the historical Buddha as its template, enlightenment in the Mahayana could be understood as a series of transformations that progressively open the nascent Bodhisattva to more expansive levels culminating in full Buddhahood. Mahayana liberation can be understood with in several different theoretical frameworks. The seeing and knowing of the *Tathagatagarba*,

Buddha-nature, or the matrix of suchness constitute enlightenment. In the Three Bodies of the Buddha model, the attainment of the highest body, the *Dharmakaya*, is becoming a Buddha. The Theravada refrain of the end of suffering and defilement is replaced by paeans to the birth of transcendental wisdom and universal compassion. A formulation that is closer to the Theravada is emptiness teaching where the recognition of the non-self-nature of all phenomena and views is the essence of awakening. The Mahayana is concerned with the various stages or levels, *bhumis*, of the Bodhisattva. In contrast to the Theravada's interest in mapping the approach to sainthood, arahantship, the Mahayana turns its attention to the open-ended career of the Bodhisattva. They both aim for what they label awakening, but they seem to trod different paths to perhaps different destinations. On the basis of classical texts, it seems that the two branches of Buddhism are speaking of different goals that use the same label and claim the same parent, the Buddha, but are in fact unrelated.

Along with the differences between the arahant and the Bodhisattva path, another apparent contrast in the view of enlightenment is the cognitive content of the experience: what is learned. The source for the Theravada view is the *Sackka Sutta* in which the Buddha takes the listener through the night of his awakening. Dividing the night into the traditional three watches, the Buddha identifies each period with the discovery of a core doctrine. The first is re-birth. Next, the Awakened One sees the law of karma, and finally the four Noble Truths are revealed. This account provides the standard Theravada version of the content of enlightenment. What is enlightenment? Stated negatively, it is the ending of defilements and suffering. Stated in the positive, it is the realization of the Eternal Dhamma, the core doctrines of the religion.

Mahayana accounts, on the other hand, say little about re-birth, karma, and the Four Noble Truths. If there is a direct seeing of a classical teaching, it is the truth of *patticasamudpada*, codependent origination, or universal emptiness. In short, there are significant differences if we are comparing maps of a path, the end statuses of attainment, and the cognitive content of the event. However, if we read for a mental process, a way of being, a quality of mind, and the very texture of the event, the gulf between Theravada and Mahayana closes. Beyond the boundary demarcations of the classic texts, the actual descriptions and personal accounts of awakening in the teachings and records of masters from both the forest and the mountains find common ground.

Before turning to first person accounts, it can be noted that there are additional scriptural sources, sometimes marginalized in commentarial literature, that indicate a proximity between Theravada/Mahayana understandings. Notably, the Pali texts do not go completely silent on the matter of enlightenment after the three watches of the night text. The Canon contains several other descriptions that approach the event from a different point of view. The Dhammadpada, for instance, contains the famed house builder metaphor in which the Buddha likens awakening to seeing the house builder, the process of ego construction, snapping the ridge pole, and breaking the critical structure of ego. He then goes on to assure the listener that the house builder "will build no more" (Roebuck 2010, p. 32). The Dhammadpada verse speaks of a critical event in consciousness that transforms, not of any doctrine that is learned. Seeing through and altering a mental dynamic sets one free. This version of awakening is more amenable to the sudden breakthrough awakenings of the mountain and forest masters.

Another Pali text that is compatible with the Ch'an and forest traditions is a short text that is held in esteem in Sri Lanka, the *Kalakarama Sutta* (Nanananda 1974). Here, the Buddha tells us how the awakened mind functions. He declares that he "does not conceive" of a self, an object of perception, or a process joining the two. Therefore, there is no basis for agitation. He repeats the same formula for each sense organ and the mind. The story of Bahiya, the bark gatherer, from The *Udana* (Ireland 1997, pp. 18–21) coincides with the Kalakarama Sutta. After being repeatedly rejecting Bahiya request for instruction, the Buddha enjoins him that "in the seeing let there be just seeing." Again, the Awakened One covers each sense channel and the mind. "Just this is the end of suffering." What is significant is that these descriptions, while less frequently called upon by Theravada teachers, match the descriptions of awakening and the awakened state by Forest Masters. In turn, as we shall see, these accounts also resonate with Ch'an/Zen.

Just as Thai forest accounts of enlightenment are often not bounded by the conventional markers of their tradition but do find support in the Pali Canon, so Ch'an/Zen is in a similar position with regard to the Mahayana world. The dizzying heights of interpenetrating Buddha worlds unfolded in the *Avatamsaka* and the *Lotus Sutra* appear to dwarf the human size *kensho/satori* (Jp.), awakening, events of today's Zen students. Nevertheless, as with the Pali/Theravada tradition, the Mahayana Sutras also contain teachings that speak to the immediate task of practice/awakening. Two core texts of the Mahayana and Ch'an, the *Heart Sutra* and the *Diamond-cutter Sutra*, deliver incisive statements that define how enlightenment is attained and how it functions. The Heart Sutra declares that "The bodhisattva holding to nothing whatever attains to *anuttara samyak sambodhi* (full perfect enlightenment)." Then how should we practice? The Diamond Sutra answers, "Develop a mind that alights nowhere" (Mitchell 1993, p. 35). Holding to nothing, abiding nowhere is both awakening and the awakened condition. The Buddha of the Kalakarama Sutta and the Udana seems to agree: Nothing the grasp, no one to grasp. No problem! According to his students, the pith instruction of Forest Master Ajahn Chah was "Let go!" Here, enlightenment is not about doctrine but about a quality of mind—the liberated mind.

Awakenings from the Forest

Whereas the Ch'an/Zen tradition has a long history of enlightenment tales, the Thai Forest Tradition offers a more limited pool of awakening stories. Their abbreviated numbers are more than compensated by the richness of communications directed to a modern audience. The following personal stories provide the data for our investigation: Ajahn Mun, Maha Bua, Ajahn Chah, Ajahn Thate, Ajahn Khao Analayo, and Mae Chee Kaew. These accounts, although sparse in number, are rich in content and exceptional in their availability. Their very accessibility is both rare and remarkable. Theravada monks are under the stricture to not share attainments with laypersons. The rule blocks monks from using claims of attainment as ways of currying favor and political prominence with the laity. Sharing within the sangha is permitted, provided, of course, the claims are honest. Given these norms, how are we in possession of these detailed narratives? How does this break occur with the usual taciturn Theravada stance?

One man sits at the center of these accounts: Maha Bua. The passages of his own awakening that appear in *Straight from the Heart (1987)*, *Arahattamagga, Arahataphala (2005a)*, and *Forest Dhamma (1995b)* are transcription of talks delivered to his monks. The letter of the law is obeyed. The talks however do not remain in a cloistered venue; subsequently they make their way into print, into English, into Thai bookshops, and online! Maha Bua moves beyond the strict customs of Pali/Theravada Buddhism with a revivalist argument. Publishing enlightenment stories will vivify the faith of Thai Buddhists in the soteriological project. These narratives sound the revivalist trumpet: The gates to the deathless are open. Come on in!

Ajahn Maha Bua Nanasampanno descriptions of awakening offer explicit detail on the process and dynamic of awakening. Perhaps only in contemporary Zen, accounts are there narratives of comparable particulars, but often without the theoretical sophistication of the Thai Master. Given his preeminent position in the Thai forest Movement as Ajahn Mun's biographer and student, and as the master of many of the third-generation forest monks, his account is prototypical. In a series of impromptu talks at his home monastery, Wat Pa Baan Taad, Maha Bua recounts the events leading to his awakening (Maha Bua 1987). While deftly evading any direct personal claim, his description leaves no doubt, that this is his entrance into arahantship. Two other Thai Forest accounts, those of Ajahn Mun and Ajahn Khao, also involve Maha Bua's hand. While as the author of their biographies, Maha Bua is technically not touting his own attainment, yet he still manages to advertise the achievements of his lineage. Maha Bua's hand is more hidden yet hovering in the background in the lively story of Mae Chee Kaew's awakening. The author, Ajahn Dick Silaratano, is a Western monk and long-term student of Maha Bua. Here, Maha Bua is not the author but a protagonist. As her teacher, he is the guide, the arbiter, and, at times, the spiritual ogre of her journey. Yet not all Thai Forest monks are comfortable with Maha Bua's promotion through awakening tales. Ajahn Jayasaro, a senior disciple of Ajahn Chah, doubts the inspirational value of these revelations, yet does not entirely repudiate Maha Bua's revivalist rationale (2008, p. 43).

Finally, the inclusion of Ajahn Chah's account requires explanation. Several of Chah's monks contend that the breakthrough experience their teacher relates is not an enlightenment moment but a samadhi state (Chah 1985, p. 86). Nevertheless, I have chosen to include Chah's story among the enlightenment narratives of the Forest Movement for several reasons. First, the content of the account is analogous to enlightenment moments of Chah's fellow monks and Mae Chee Kaew. There is a phenomenological match. Samadhi states are classically presented as absorptions and classified by degrees of one-pointedness. In contrast, Chah's experience is one of explosive dis-identification. Next, Chah follows up this dramatic event with a discussion of the profound shift in perception, motivation, and affect that can occur on the path. While he does not explicitly link the two, it is reasonable to conclude that he is implying a connection simply on the basis of their sequence in the discourse. These shifts Chah notes are significant maturations of the path. Was the altered state of consciousness enlightenment? Perhaps not, but it is more than just concentration.

Rather it has an awakening quality and function. Chah has a subsequent experience of marked detachment that Jayasaro agrees is an awakening (Jayasaro 2008, p. 29). However, if we could query Ajahn Chah, he would probably say, "Fish sauce."

From the Mountains

If direct pointing and presumably seeing is the signature of Zen, then it is surprising that classical Zen is so sparse with regard to phenomenological accounts. Zen koans, those dramatic and non-rational enlightenment stories, usually limit themselves to "And then the monk was enlightened." We are told about the foreplay but not about the climax itself. These abbreviated accounts are corrected by more revealing descriptions from later Song/Ming dynasty documents and by contemporary reports. Our sample draws on autobiographical documents from post-T'ang Dynasty China. One of the earliest extended personal reports of awakening in the Ch'an school is Han Shan (b. 1545). Ordaining at a young age, he had intimations of the great awakening to come in his experience of the stillness of all phenomena that he interpreted as the teaching of "nothing moves or arises." Years later, while practicing in his mountain hermitage, he awoke to what he called the "Illuminating Whole." He describes his realization of an "omnipresent, perfect, lucid, and serene" truth. He tells us that he had "no body or mind." His relationship with the external world is transformed: "sound, voices, visions, scenes, forms, and objects were no longer hindrances." He had entered into a world of transparency where "All my former doubts were nothing." In Han Shan's account, we encounter several of the major themes of awakening: suddenness, radical perceptual change, lucidity, and certainty (Chang 1959, pp. 118–142).

From Japan and the seventeenth/eighteenth century comes the remarkably revealing autobiography, *Wild Ivy (Itsunadegusa)* of Hakuin, the reformer of the Rinzai School. A contemporary Ch'an account appears in the writings of Sheng Yen. This text is supplemented by personal stories from two of Ch'an Master Sheng Yen successors with whom the author has had personal contact: Guo Jun, formerly the head monk at Pine Bush retreat center and presently an Abbott at a Singapore temple, and Guo Gu (Jimmy Wu) who heads the Tallahassee Ch'an Center and is Buddhist scholar at Florida State University. Both describe enlightenment in audio/video recordings. Sheng Yen's book *Getting the Buddha Mind* also contains a number of first-hand retreat accounts some of which contain the Master's approval of students' realization.

The modern Japanese contribution comes largely but not entirely from the Sanbo Kyodan, a twentieth century movement initiated by Zen monks to revive what they saw as a moribund Zen establishment. Their practice is lay-accessible and enlightenment focused. Members are encouraged to publish their experiences and are held up by the movement for public recognition. Besides being in contrast to Theravada reticence, the movement's fidelity to classical Zen has been questioned by Sharf who label it a New Japanese Religion (1995). Nevertheless, it must be noted that many of the prominent Zen centers of North American count Sambo Kyodan teachers among their lineage ancestors. Much of Western Zen is Sanbo Kyodan Zen. In addition, the movement/cult is undeniably revivalist, making it pertinent for this study. Accounts of Sanbo Kyodan disciples have been compiled into two volumes by Yasutani Roshi. A third book is edited by Abbess Sozen Nagasawa (Dumoulin 2006, p. 127). Selections have made their way into Kapleau's *Three Pillars of Zen* (1965). Adding to these reports of contemporary Japanese-based experiences are Zen historian Heinrich Dumoulin's work on Zen enlightenment (2007) and K. Sekida's (1975) piece on Zen training. More recently, R. Forman's (1999) study of mysticism has included an interview with a Western Roshi on the matter. Taken together, we have both contemporary and classical renderings from Chinese, Japanese, and Western students. While hardly constituting either a complete or statistically random sample, they are a collection of cases which permit cautious generalizations.

Dimensions of Awakening: Phenomenology, Structure, and Function

Accounts of enlightenment experiences offer opportunity for analysis and comparison from several perspectives. The first is the descriptive. What was the experience like for the actor? What are the tones and textures of consciousness in awakening and do they compare with one another or are we confronted with completely unique personal moments that defy generalization? The phenomenological aspects of Thai Forest narratives can be sorted for recurrent themes and metaphors which in turn can be matched with Ch'an/Zen accounts.

It might be objected that experiences claiming to be beyond language cannot be understood from language-based accounts. First, resorting to language after the event does not mean it was present during the event, it may indeed be beyond words, with words offering only a groping, ex post facto approximation. As one of my Zen teachers noted, Dogen, the great thirteenth century Japanese teacher, knew that ultimately language falls short, yet because we are human, homo-linguae, so we try. It is the best we have. Second, some might downgrade these accounts to merely a repetition of stock phrases that cannot affirm any commonality (or difference). This fails to appreciate the absence of communication between the forest and mountain, and for all the limitations of language, the speech of the masters is fresh and personal and draws from no shared catechism. They do not simply parrot the suttas and sutras.

The second dimension to be explored is the structure and dynamic of the mind in and after awakening. The available narratives not only reveal an experiential event but also suggest structural transformations of the mind. Comparisons, analogies, and metaphors tell us more than what it "felt like" they point to what happened. What, at the level of mind, not transient feeling, changed? The distinction is between "there was an immense peace" and "the mind seemed to expand to infinity." One statement gives an affective tone; the other denotes a change in constitution. Here, we go beyond momentary experience to a transformation of constituents of knowing.

Third is how the event shapes the movement's social relations and its place in a broader web of meanings. What is the function of enlightenment for the social movement? Of course, the differences between phenomenological description, structural forms, and functional consequences are not absolute. Statements can be analyzed from multiple perspectives. Nonetheless, each domain will be examined separately.

The following categories emerge from a phenomenological examination:

1. Collapse and expansion
2. Dropping body and mind
3. Extraordinary yet ordinary
4. Imagelessness
5. Suddenness and immediacy
6. Rapture and light

A structural analysis suggests the following:

1. Non-duality
2. Dis-identification
3. Participatory knowing

4. Inversion of consciousness
5. Ultimate Reality: Mind

The broader social functions of the enlightenment moment suggest three categories:

1. Inter-personal verification
2. Charismatic authority
3. Sacred Canopy

The Phenomenology of Awakening

The Theravada Buddhism with its strong textual tradition possesses a robust lexicon to label and locate enlightenment within a complex and systematized teaching. Dhamma talks, when they on rare occasion turn to the subject of Awakening and the Unconditioned, situate liberation within conceptualizations of the path and components of practice. Furthermore, we are told of varieties of enlightenment, for instance with knowledge (conceptual understanding) as opposed to wisdom (intuitive awareness). Awakening is related to the factors that facilitate it. Yet what is actually experienced by the actor? What is it like as a subjectively lived event? Classical texts, scholars, and teachers are notably silent. Only in a practice lineage, such as the Forest Tradition whose authority derives from the enlightenment event, do we encounter first person narratives conveying the heart of awakening.

1. Collapse and Expansion

   A recurrent metaphor of awakening is one of collapse, breakthrough, and expansion. A structure, the sky, the earth, the bottom of a bucket breaking open results in an expansion of awareness, sometimes to the end of the universe. A hardened or at least limiting form is broken open releasing what it had enclosed, for example, a foot in a shoe, water in a bucket. Frequently, this entails an implosion/explosion of consciousness.

   Ch'an master Wu Wen's awakening is signaled by his body and mind sinking "like a house whose four walls had fallen" (Chang 1959, p. 143). Hsueh Yen tells us that "something broke abruptly before my face as if the ground were sinking away" (p. 147). A contemporary audio report by Gou Jun (2013), a senior monk of Sheng Yen, describes a cracking open of an invisible barrier and his subsequent tears with the end of dualistic isolation. In the forest tradition as in Ch'an/Zen, a collapse and concomitantly an expansion occur. Maha Bua offers a simile of awakening: "like a water jar whose bottom has been smashed" (1987, p. 225). Kicking out the bottom of a bucket is a Zen favorite. Ajahn Khao Analayo refers to a "world shaking event when the sky and the ground collapsed" (Maha Bua, p. 90). We are told that Mae Chee Kaew "felt her being dissolve, expand outward and merge with all things" (Silaratano 2009, p. 186). For Maha Bua, "the universe collapsed releasing consciousness to cover the entire universe" (Maha Bua 1987, p. 235).

2. Dropping Body and Mind

Concurrent with a collapse or breakthrough are reports of the dropping off of body and mind. In both traditions, the moment of realization entails a radical change in relationship with the body. The body disappears or shatters. It is no longer the familiar reference point of separate individual identity. In the Ch'an/Zen school, this is dubbed the dropping of body and mind. Dogen credits his Chinese teacher with the phrase and its sharp utterance (accompanied by a slap from a sandal) for his own Great Awakening. *Shinjin dasuraku* (Jp.), dropping off of body and mind, is Dogen's battle cry of awakening. While this exact phrase is absent in the Thai Forest Tradition, the experience is attested to repeatedly.

The dropping of body consciousness or its radical attenuation is so common that one is hard pressed to find a Ch'an/Zen account that does not mention this phenomenon. K. Sekida, a Zen scholar and practitioner, uses the term the "off sensation" to designate this loss of body awareness. In Forman's (1999, pp. 19–24) interview with John Daido Lori Roshi, the latter describes a complete absence of body consciousness interrupted by the brush of his teacher's robe as he passed by. Suddenly, he came to awakening. A traditional text that attests to the loss of body consciousness in awakening is Hakuin's autobiography. In one instance, he is knocked out cold by an angry old woman wielding a broom. Zen awakening stated positively is to know one's true nature. Stated negatively, it is, according to Dogen and the tradition, the dropping off of body and mind of self and others. Hakuin uses the same phrase in describing his great awakening (Waddell 2011, loc. 1053).

Chah details a process of his mind turning inward repeatedly. On the second turn, his "body broke into fine pieces" (Chah 1985, p. 185). Mae Chee Kaew dissolves into the universe (Silaratano

2009, p. 199). The spiritual warrior Khao Analayo describes "seeing the parts within his body quite clearly," allowing him to de-construct the body sense (Maha Bua 2006, p. 77).

3. Extraordinary Yet Ordinary

   The drama of the moment is paradoxically accompanied by a reflection on its ordinariness. The extraordinary ushers in the ordinary. Everything is transformed, yet is just the way it always has been. The line from T.S. Eliot's *Four Quartets* (1968, p. 59) comes to mind, "We shall never cease from exploring only to return to the place from which we began, but to know it for the first time." Instead of exotic realms, the enlightened mind sees a woman walking on a country road, an old tea pot in the kitchen. All is utterly natural, ordinary, but for the first time.

   Te-Shan said, "what is known as 'realizing the mystery' is nothing more than breaking through to grab an ordinary person's life" (Buckley-Houston n.d). After being repeatedly probed by Master Sheng Yen on the identity of who was answering the kung-an a monk exploded into the immediate and the ordinary, "I am Chi-Ch'eng. Chi-Cheng is me." Another retreatant after his breakthrough notes the birds were singing outside. "Buddhism is really empty, isn't it?" The pupil and master laugh (Sheng Yen 1982, p. 127, p. 92).

   Maha Bua after his attainment of arahattaphala with its brilliant citta and consciousness that blankets the universe observes that "everything still exists as it always has" (1987, p. 236). Perceptions of simple, prosaic village scenes are utterly compatible with deep meditative awareness for Ajahn Chah. Although everything is different, Chah realizes that "it was really just I who had changed, and yet still I was the same person"(Chah 1985, p. 186). Elsewhere, he concludes, "There's nothing more to it than this." "This is called lokavidu—knowing the world clearly as it is" (2002, p. 256). Jayasaro, an eminent disciple, considers this event to be an enlightenment attainment by Chah (2008, p. 29).

4. Imagelessness

   A corollary of the ordinary is that awakening in both these schools is not image dependent. There is no content of consciousness, image, sound, and sensation that signifies enlightenment. This type of mysticism has been called Pure Consciousness Experience (PCE), and is contrasted with religious experiences that center on a religious form, Jesus, Krishna, or Buddha (Forman 1999, p. 11).

Awakening is a not-image-dependent event even as sensory experience may be present. For this reason, not-image-dependent (NID) is perhaps a better designation than PCE. Also, it should be noted that this type of experience is not unique to the two movements under investigation. Meister Eckhart and Ibin Arabi are representative of NID awakenings in the Christian and Muslim tradition, respectively. For Ch'an/Zen, the NID quality of the event is essential. Awakening goes beyond the world of form, *rupa*, even religiously sanctioned forms. In the Thai Forest Tradition and the Ch'an/Zen school, there is no place for visions of Buddhas and Bodhisattvas, ghosts and devas.

Mae Chee Kaew is severely chastised for her flirtation with spirits. Maha Bua describes the awakened mind as utterly clear with all images having disappeared and returned to their source point. "After the internal images had all disappeared, the citta was empty" (1987, p. 209). The use of terms like "clear," "empty," and "imageless" in both traditions indicates that awakening is not-image dependent. Not a single account reports an image, audio phenomena, or kinesthetic sensation as evidence of awakening. The transformation lies in the quality of the knowing not what is known. What is known, as we have shown, may be quite prosaic, but how it is known is transfigured.

5. Suddenness and Immediacy

In the Ch'an school, the word "sudden" and the word "enlightenment" are inseparable. Ch'an branded itself as the sudden school and claimed a distinction and superiority over its competitors. A qualification on this position was offered by Korea's great Son Master, Chinul, who spoke of sudden enlightenment and gradual cultivation (Buswell 1992, p. 101). Hakuin also speaks of post-satori practice. After awakening, there is slow work to be done. Today in both Japanese and Chinese schools, it is typically taught that initial awakenings need to be integrated and followed by additional openings to the nature of Mind.

Official Theravada teachings, in contrast, insist on enlightenment's gradual nature and in support of their view cite passages delineating the progressive approach to realization. Both Ch'an and Thai Forest schools therefore have a heritage of opposing official positions on the question. However, as we will show, this is more product advertising than what is in the can. There is a different view on the run up to enlightenment, practice, and its role, but in regards to the experience itself, both schools agree it is sudden, immediate, and a distinctive. Awakening is not a slow motion event.

As we would anticipate, Ch'an/Zen accounts are replete with words like "abruptly" and "suddenly." "Instantly my mind became frozen," relates Hsueh Yen. The experience is like "bright sun shining suddenly into a dark room" (Chang 1959, p. 148). Meng Shan's tells us, "suddenly and abruptly I recognized myself" (p. 151). Wu Wen recounts, "Suddenly I felt my mind become bright, void, light, and transparent"(p. 143). The moment is always quick and decisive.

What is perhaps less recognized is that the phenomenological reports of awakening in mountain and forest agree that the event is qualitative and unequivocal. Maha Bua assures us that the disintegration of ignorance is instantaneous. "When it disbands, it disintegrates in an instant, like lightning flash" (Maha Bua 1987, p. 224). Awakening is so instantaneous and temporally precise that Maha Bua can give the exact hour of Ajahn Mun's consummation of the path (Maha Bua 1995a, p. 128). Ajahn Silaratano gives the date of liberation for Mae Chee Kaew (2009, p. 199). For Ajahn Khao, the universe collapses of itself. Whatever the Theravada understanding of the gradual coalescence of factors that lead to awakening, the event itself, according to the Forest Tradition, is sudden.

6. Rapture and Light

Awakening is an intensely affective and sensory experience. No matter how coolly cognitive Buddhism fancies itself, awakening is an emotional moment. Certainly, Buddhist teachers are suspicious of displays and hyperbola. *Hoetsu* is a Japanese term translated as Dharma-delight and is a common component of Zen awakenings. Yasutani reminds us, "Joy and enlightenment are two different things" (Dumoulin 2007, p. 144). Nevertheless, they seem to consistently accompany one another. A modern Zen student exclaims, "I was astonished that unnoticeably the Zen Hall and I myself were radiant in absolute light" (p. 145). Han Shan tells us, "My joy is so great, words cannot describe it." Streams of tears fall from Hakuin's eyes (Dumoulin 2007, p. 151). Forest teachers testify to rapture and light as well. Maha Bua weeps with joy. Khao Analayo is illumined by a light seemingly coming from the sky (Maha Bua 2006, p. 77). While neither rapture nor light are enlightenment, they are undeniably part of the phenomenology of the moment and a shared feature in both Ch'an/Zen and Kammatthana experience.

The Transformation of Structures of Mind

Descriptive statements imply structure and function. Phenomenological reports suggest changes in the pattern of perceiving and the organization of the mind. Experience happens but its import is that the subject is transformed. The transformation entails perceptual shifts that go beyond phenomenological descriptions. The following four changes will be explored: non-duality, dis-identification, participatory knowing, and the inversion of consciousness.

1. Non-Duality/Unity

Non-duality, the view that ultimately reality is not-two, has been identified as a hallmark of Asian thought (Loy 1997). Certainly, there are other philosophical positions on the continent, but not-two-ness has a central place in schools of Buddhism, Hinduism, and Taoism. The nondual has made it appearance in the West as well, but it has been greeted with skepticism and marginalized. Here, our concern is with the enlightenment moment itself rather than philosophy. Enlightenment is a nondual, unitary consciousness moment for both Ch'an/Zen and The Forest Tradition.

The claim of unity is more prominent in the Ch'an/Zen school. In fact, it may be the hallmark of awakening in the modern Sanbo Kyodan sect where students are ecstatically unified with fences, watches, tatami mats, and indeed the whole universe (Kapleau 1965; Sekida 1975). "Mu" is you, me, and the cat seems to be the central insight. Historical Ch'an/Zen also contains the non-dual claim. Dogen shouts that the mountains and rivers are Buddha. Han Shan sees a pellucid Illuminating Whole. All phenomena, cognitive and physical, arise out of the One Mind (Chang 1959, p. 139). Sengtsan, the Third Ancestor of Ch'an/Zen, in his classic poem, *Faith in Mind*, offers the universal elixir to delusion. "When doubts arise simply say: not two." He concludes his verse with "To live in this faith is the road to non-duality because the nondual is one with the trusting mind" (Clarke 1973, p. 12). Dongshan, a founding father of the Caodong/Soto school, awakens upon seeing his image in a stream. "Everywhere I look it is me," he declares before intriguingly adding, "but I am not it" (Leighton 2015, p. 34). Zen assiduously avoids asserting that everything is One, while equally insistent that it is not-two. Reality is non-dual.

The theme of unity is less central and consistent in forest accounts, but not entirely absent. Maha Bua's consciousness blankets the universe. Mae Chee Kaew's melts into the cosmos. Both are undoubtedly unitary experiences. Maha Bua offers further evidence beyond ecstatic utterances in his insistence on the annihilation of the "owner" as essential for authentic awakening. What Maha Bua terms the "owner" is the observer-self; that structure/process whereby there is a persistent watcher or "owner" of the content of consciousness. "If there is a point or centre of the knower anywhere" that is ignorance (Maha Bua 1987, p. 230). To collapse the knower is to enter into non-dual consciousness. This understanding is not to be found in mainstream Theravada characterizations of enlightenment. If non-duality is implied in Theravada teaching in Maha Bua's exposition, it is explicit.

2. Dis-Identification

   Awakening is a radical dis-identification with the usual reference points in consciousness. Perhaps the most significant referent, for which we have already seen evidence of dis-identification, is the collapse of the body and the mind. In awakening that one is neither *rupa*, form, nor *nama*, mind, ceases to be mere theory. The not-two-ness of the moment already implies that there is no one to be or exist as. If the corporeal and the mental operations are not you, then it follows that identification with gender, age, nationality, community, and kinship offers no place to abide.

   After recounting a non-ordinary experience, Chah discourses on its transformative effect. "One's whole world has turned upside down, one's understanding of reality is different, people appear different, everything changes, thoughts are transmuted so one thinks and speak differently than others, and one is no longer the same as other human beings" (Chah 1985, pp. 183–186). Han Shan expresses this transformed relationship with the world in poetry: "I can look back at the world again, which is filled with nothing but dreams." He tells us, "Sounds, voices, visions, scenes, forms, and objects were no longer hindrances." For Hsueh Yen, "The whole day seemed like a passing second. I did not even hear the sound of bells and drums, which occurred at regular intervals in the monastery." This compares with Chah's report of hearing but being able to shut out at will the sounds of the nearby village. Master Kao Feng describes his states as that of a "living corpse." He is in but not of the world (Chang 1959, pp. 131–153).

3. Participatory Knowing

If there is no two-ness and if one is a dis-identified with the contents of consciousness, then how does one know? What is the epistemological mode of the enlightened state? At this point, the mystic often demurs from further explanation. However, an investigation of the enlightenment reports and the interpretations of interested scholars suggest that the awakened Ch'an monk or Forest bhikkhu does not know "of" or "a" but "as." We can call this knowing "participatory" in that knowledge arises by being it what is known. Colloquially, we can say it is a knowing from the inside out. This kind of knowing can occur outside the enlightened state, and is the inner disposition of all Zen outer arts from the martial to flower arranging. Participatory knowing is the secrete of kung-an (koan, Jp.) practice. The student must answer without a trace of separation. Spontaneous participation is demanded. He who hesitates is lost.

Awakening in the Forest Tradition and Ch'an/Zen is less fixed on what is known and instead underscores this transformed mode of knowing. The de-construction of the separate knower untethers awareness from any observer stance and allows the emergence of a new way of knowing. Mae Chee Kaew's experience as worded by Ajahn Dick states it perfectly.

> The knower was everywhere, but nothing was known. Without characteristics and without source, emanating from no point in particular, knowing was simply a spontaneous happening of cosmic expanse (Silaratano 2009, p. 199).

Turning again to Dongshan his awakening exclamation is an expression of participatory knowing. Seeing his image in the stream is the opening to seeing himself everywhere. The universe is Dongshan's mirror. He is no longer a discrete subject standing outside. He is his perceptions. He knows by being what he perceives—in other words participatory knowing.

A theoretical reflection by Ajahn Chah provides an understanding of this transformed mode of perception. Chah asserts wisdom and samadhi to be two sides of the same coin. Samadhi, a non-dual awareness, is the passive side, while wisdom, panna, is the active (2002, p. 249). Taken together, this suggests that panna is a mode of knowing in "at-one-ment" or participatory knowing. The implication is further strengthened by Chah's teaching tactics which as explored in Chap. 2 sought to force the student into unreflective action of total involvement. Finally, Chah's notion of samadhi and panna as inexorably connected aligns with the revolutionary teaching of the Sixth Ancestor of Zen who expounds in his *Platform Sutra* a doctrine of wisdom, emptiness, and an open-eyed samadhi as the multi-faceted dynamic a single reality (Red Pine 2006, p. 10).

4. Inversion of Consciousness

   Personal experiences and scriptural guidance indicate another structural alteration: the inversion of consciousness. By inversion is meant a radical reorientation of consciousness not in what it perceives but in its direction. The normal directionality of consciousness, from subject to object, is turned inside out, usually by turning attention on its apparent source. This maneuver is obviously related to other structural markers such as non-duality, participatory knowing, and dis-identification. The Ch'an/Zen tradition can trace this transforming maneuver back to the Lankavatara Sutra's declaration that enlightenment is "a turning about in the deepest seat of consciousness" (Suzuki 1978, p. 125). Significantly, the Lankavatara Sutra was the premier text of the early Ch'an movement. Bodhidharma and his successors were "Lankans" whose insistence on knowing the mind itself put into practice the Yogacara, mind-only school of Mahayana thought, contained in the Lankavatara Sutra. Throughout the 1500-year history of Ch'an/Zen, outstanding teachers have counseled the turning the light of consciousness on itself. Likewise, the Thai Forest Tradition has preached the turning of consciousness on itself or looking into the origin point of the light. For both movements, entering the gate of awakening is less about what is seen than about the direction of the looking.

   The idea that awakening entails a turning about of conscious find testimony in the already mentioned "seeing the house builder" phrase of the Buddha. By turning back on consciousness itself, one is liberated from the mind's magic show. T'ang dynasty Ch'an master, Shitou, in his vibrantly personal poem, *The Grass Hut*, tells us to turn, "and then just return," dropping off eons of anxiety. Hongzhi, the Song Dynasty Master and poet of the awakened mind, counsels "when you turn within and drop off everything completely, realization occurs" (Leighton 2000, p. 19). His grandnephew who transmitted Caodong Ch'an to Japan, Dogen, called turning the light within the essential art of zazen. He repeatedly urges his monks to take the backward step, *ekehensho* (Jp.) (Cleary 1986, p. 9). Today Jayakuso Kwong, a Chinese American Zen teacher, advocates turning the mind on the "sourceless source" of consciousness (Kwong 2003, p. 240). We find this practice of directly entering enlightenment being taught by arguably the outstanding Ch'an Master of the twentieth century, Hsu Yun. Master Tsung Kao even contends that laypersons are capable of making a "much more thorough and mighty *turnabout* than the monk" (Italics original) (Chang 1959, p. 90). Hsueh Yen tells us, "One day I suddenly turned my mind inwards," searching out "the point where thoughts arise and disappear. It became serene, and limpid" (Chang 1959, p. 145). A full review of this practice and its varieties in different Zen schools is beyond the scope of this study; yet clearly the inversion of the mind on its self is a recurrent meditation instruction in Ch'an/Zen.

If this is the essential art of zazen explicitly taught by the Ch'an/Zen school, then it is for the forest monk a discovered skill. Practicing alone without his recently deceased teacher, Maha Bua faces a conundrum. His mind is radiant, illumined, yet he intuits something is amiss. He is being deceived by a light more seductive than Mara's daughters. By turning awareness back on the point/source of the radiance, he comes to see the lair of delusion, the deception of a knower. The key is the inversion of mind on itself. "If we don't look back at ourselves, we won't see ourselves," Maha Bua concludes (1987, p. 224). We must, in Dogen's words, "turn the light within." Ajahn Sumedho also promotes an inversion of consciousness either by sound or by reflecting on the knowing factor in the moment. To what extent his early practice of the Ch'an technique of knowing the knower influenced his later approach is not certain.

5. Ultimate Reality: Mind

As the awakening of Ch'an/Zen and the Forest Masters is not image dependent and if it entails a turning of consciousness of itself, then what is known? Our review of mainstream Theravada thought identified enlightenment as disclosing the doctrines of Karma and Rebirth, the Four Noble Truths, and the Eightfold Path, just as the Buddha reported from his three night watches. However, references to these core doctrines are minimal at best in the Thai Forest Tradition. Rather than specific doctrines, Thai Forest Masters speak of seeing the Dhamma. Here, Dhamma means Ultimate Truth, what the Buddha in several Pali passages calls the shining Dhamma, the Deathless Dhamma. In a language strongly reminiscent of Ch'an, the Forest Tradition also speaks of the pure or true nature of mind as what is realized. This is our true nature, natural mind, radiant mind, or simply the heart. If we turn on the source of mind, then it is the very nature of the mind, transcending duality and content, is realized.

True to its founder, Bodhidharma, Zen points to Mind or Reality itself. Bodhidharma's offspring speak of their practice as looking into the nature of mind, meditating on mind itself, or in T. Cleary's (1995) phrase "minding mind." The very term for awakening in Japanese, ***kensho***, literarily means to see, ***ken***, into the fundamental principle, ***sho***. Drawing on Chinese philosophy that speaks of function and principle, enlightenment is seeing the fundamental principle of Reality, which for a Buddhist, is Dharma. Here, the languages of Ch'an and Forest Masters converge. They talk of the "true," "original," "pure," and "natural" mind or heart rather than any particular doctrines. Core teachings are not repudiated, indeed they are affirmed, but not as the essential discovery. As Maha Bua has attested there is no owner of pure mind. Not self, anatta, is a necessary but not the ultimate insight. Boldly, Maha Bua goes further and speaks of a truth in addition to the Four Noble Truths, pure mind. "It's a truth beyond the Four Noble Truths," he declares (Maha Bua 1987, p. 226). For both schools, enlightenment is the entrance into reality, and that reality is pure mind. This point will be further examined when we turn to the doctrines of Ch'an/Zen and the Thai forest tradition in Chap. 7, 8.

The Social Functions of Awakening

Lastly, what is the place of enlightenment in the worldview, the authority patterns, and social interactions of the Ch'an mountain masters and the Thai Forest sages? Moving beyond the experience, what is the meaning assigned and how does it shape community? Changing from a microview to a macro-overview of the functions of enlightenment, we discover that the Ch'an/Zen and Thai forest movement mirror one another in the central position enlightenment plays in their webs of meaning. Here, we examine the explicitly social dimensions of awakening. Our three areas move from the inter-personal, to patterns of authority, to the symbolic superstructure of the two movements.

1. Inter-Personal Verification

Critical to the dissemination of enlightenment into the social world is its verification. The confirmation of enlightenment discloses much about its larger role in both schools. The question of verification brings awakening out of the private zone into the inter-personal and the collective. What might under other conditions simply be someone's odd moment becomes a public identity that confirms and extends the purpose of the entire school. The manner of the verification also reveals the locus and the transmission of the sacred and in turn the authority it confers. Verification returns us to the heart of the teacher/disciple relationship. If the teacher's role is to guide the student toward liberation, then that performance is consummated in the affirmation of ultimate attainment by the student.

Verification of realization is a face-to-face encounter between student and teacher. In Zen/Ch'an, this is called "mind to mind transmission." This phrase has led to unfortunate misunderstandings. The notion of a transmission between minds conjures up a grade B sci-fi movie with light rays and sound effects. Rather the process as has been confirmed by several teachers is a mutual recognition of a shared mind state. What is of a quasi-mystical nature is that student and teacher meet one another in an identical condition, but this is not unlike any intimately shared moment. What is transmitted is affirmation of an opening beyond the conditional, not conveyance of a supernatural substance.

Non-verbal communication plays an important role as the master synthesizes and interprets clues about the disciple's inner state. However, as we shall show, the verbal is not absent or unimportant. The student typically reports her experience or dramatically demonstrates it. A cross-examination may ensue. Finally, although Buddhists reject the idea of transmitting enlightenment (in Buddhism no one can enlighten another), there is a supernatural power often believed to be at play in this meeting. Among the supernatural talents (*iddhis*) recognized in Buddhism is reading another's mind state. This is referred to as "looking into" the other's mind. The encounter where enlightenment is verified (or rejected) brings together a variety of modes of communication and knowing. While the euphemism having a "dhamma conversation" is used by forest monks for these verification encounters, closer questioning of informants revealed a belief that the mind of the other was being directly perceived by the master.

Historical evidence modifies this ideal model of verification. Research suggests that certain Zen masters of the Song and Ming dynasty may not have been passed by representatives of any known lineage. The problem proliferates in North American Zen with self-anointed roshis who never completed their studies or were never empowered to teach. Japan has its own lapses as there are apparently certifications that entail no face-to-face encounter much like degrees granted on the basis of self-graded take home exams! Nevertheless, the model is presented to the world as the heart of Zen practice and is adhered to as much as possible. All Ch'an/Zen retreats attended by the researcher featured almost daily interviews with the presiding teacher.

Roshi Glassman after his great breakthrough realized he had to excuse himself from work to seek out his teacher Maezumi for confirmation. Guo Jun awaited Sheng Yen's arrival at Pine Bush Retreat center to present his experience. Maha Bua initially rejected Mae Chee Kaew's claim pointing out a subtle fluctuations in her mind. Verification of liberation also extends beyond the Forest Tradition proper. Upon meeting Kee Nanayon, an independent female adept, Maha Bua examined her mind and confirmed her attainment, according to Pra Panlop who accompanied his Master to her retreat center.

The verification process is inter-personal but not necessarily one way. Stories abound of students testing teachers. Indeed, Zen koans often entail a vigorous back and forth. Bankie, a Rinzai monk and latter national teacher, rejected teacher after teacher, both Japanese and Chinese. Dogen exhausted the teachers in Japan went to China and only after a long search did he find Ju-ching, a true living Buddha with whom he could study.

The relational aspect of enlightenment also implies a socially defined context. All the actors have agreed to the rules and the definition of the situation. Even the most outlandish behavior implicitly relies on social context. This is not to reduce enlightenment to a social construction but rather to acknowledge the role of context. Seung Sahn, a contemporary Korean teacher, displayed his Soen/

Zen freedom during an intensive retreat by mischievously rearranging monk's sandals. This was taken as a possible sign of his breakthrough (1976, p. 229). The meaning of behavior, no matter how unconventional, is determined by the definition of the situation: monk, enlightenment, meditation retreat. Clearly, not all sandal thieves in Korea are presumed to be nascent Zen masters.

While the verification moment is dramatically intimate, a timeless event of awakened mind meeting itself, there is a context of relationship, community, and history. There are cases of individuals who have never formally practiced approaching a teacher, but this is uncommon in the annals. Usually, there is an established relationship or at least apprenticeship with teachers within the tradition. Context both makes collective sense of the experience and translates it onto a larger stage.

2. Charismatic Authority

Two classical concepts from the social sciences can clarify the dynamics at play in verification of awakening. First is the "sacred," as delineated by Durkheim. Second is Weber's ideal type, "charisma." Although Durkheim has been taken to task for applying universally a concept that may be a Western religious notion, it has some explanatory power in Buddhism. Buddhist dichotomies such as samsara/nirvana and sanghata/asanghata are rough parallels to the profane/sacred dichotomy. What characterizes the sacred, which can be good or evil for Durkheim, is its power beyond the conventional world and that it is marked off by symbol, ritual, and prescribed behavior from the profane.

In Thai Buddhism, deference is displayed toward certain objects, Buddha images, and places and objects of power, but in the world of the peripatetic Forest monk, it is the monk himself who carries the sacred. The enlightened arahant is a vessel of the sacred. He has stepped through the Door of the Deathless and embodies the beyond. The class of sacred objects such as Buddha images is designated in the Thai language as *ong*. Monks are also classified as ong, sacred objects. The much noted prominence of relic worship in the Theravada further reveals the locus of the sacred as the monk himself. Of course, the holiest of holies is the enlightened monk. Enlightenment, embodying the sacred, confers the authority Weber termed "charisma."

The charisma of the Ch'an master or the kruba ajahn of the forest movement is not the absolute charisma of Max Weber's ideal type. The teacher is not free of all norms and the initiator of a new religion guided only by his own authority as was the Buddha. Yet he does rise to the level of the authoritative mediator of truth. They do not replace the Buddha but become his embodied representative. Lord Buddha is gone but his agent, the master/teacher, is vividly present. This claim finds support in the Buddha's pronouncement that Dharma is "known by the wise." The wise teachers of the forest and mountains know for themselves and therefore can guide and confirm others. Enlightenment and its most vivid display, transmission, both affirms authority and strings together social relationships within the group and its imagined ancestors.

3. Sacred Canopy

Religious communities, even entire civilizations in the past, are cloaked in what Peter Berger called a sacred canopy of meaning (1969). Community members are shielded from the unknowable, the meaningless, and the chaotic by a cloth that covered their lives with meaning. However solitary the early brothers of the mountains and the forests, they remained members of a religious community or at least tradition. Their very purpose, to realize and transmit the Buddha's dharma, makes sense only in relationship to a shared tapestry of meanings. What is the place of enlightenment within the woof and wharf of the sacred canopy of the Thai Forest Movement and Ch'an/Zen?

As we have seen enlightenment is confirmed in a social interaction and provides the vital source for the authority structure of both movements. In addition, the design of their canopy is organized around a central image, awakening. As revivalist movements, they can draw upon the core of Buddhist doctrine to assert not only the centrality but the availability of liberation for the Buddhist project. Here, the scriptures support both Ch'an/Zen and the Forest Tradition. The Buddha proclaims that like the great ocean, his lengthy and complex teachings have one salty taste, freedom (Ireland 1997, p. 70). The Buddha's assertion of the singular goal of his mission is reaffirmed by Bodhidharma's direct pointing and Ajahn Mun's unwavering pursuit.

Summary

Our review of enlightenment in the Forest and Ch'an/Zen movements identified similarities at the levels of descriptive content, transformed structures, and social functions. Once we move past doctrinal discussion based on selected classical scripture, we are confronted with the likeness between the two movements. A phenomenological review reveals similar metaphors of breakthrough to a NID state entailing a dropping of body, a paradoxical ordinariness, a collapse of conventional reality, and a direct knowing of ultimate reality accompanied by experiences of rapture and light. The inversion of mind, non-duality, and radical dis-identification are structural shifts. Mind is reported to be reorganized in similar ways in both schools.

The social role of enlightenment from its face-to-face verification to its bestowing of charismatic legitimacy was examined. While the Theravada concern with purity and its four-stage enlightenment model differs from Ch'an/Zen, the phenomenology of awakening, and its mental transformations, and social functions are remarkably similar. The assertion by each of these schools that they are reviving Buddhism rests largely on their claim as transmitters of awakening. Here is not only the fulfillment of their soteriological project, liberation, but also a connection to the root charismatic, Lord Buddha. Through their enlightenment, the Buddha and Buddhism continue.

References

Berger, P. (1969). *The sacred canopy*. New York: Doubleday Anchor.

Bua, M. (1987). *Straight from the heart*. Udon Thani: Forest Dhamma Books.

Bua, M. (1995a). *The venerable Phra Acariya Mun*. Samudra Sakorn: Wat Prajayarangsi.

Bua, M. (1995b). *Forest Dhamma*. Udon Thani: Forest Dhamma Books.

Bua, M. (2005a). *Arahattamagga, Arahattaphala*. Udon Thani: Forest Dhamma Books.

Bua, M. (2006). *Venerable Ajaan Khao Analayo*. Udon Thani: Forest Dhamma Books.

Buckley-Houston, T. (n.d.). Being ordinary. Retreived April 4, 2015, from http://www.buddhistgeeks.com/2010/03/being-ordinary/

Buswell, R. (1992). *Zen monastic experience*. Princeton, NJ: Princeton University Press.

Chah, A. (1985). *Still forest pool*. Wheaton, IL: Theosophical Publishing House.

Chah, A. (2002). *Food for the heart*. Boston, MA: Wisdom.
Chang, G. C. C. (1959). *The practice of Zen*. New York: Harper and Row Publishers.
Clarke, R. (1973). *Faith in mind*. Buffalo, NY: White Pine.
Cleary, T. (1986). *Shobogenzo: Zen essays by Dogen*. Honolulu, HI: University of Hawaii Press.
Cleary, T. (1995). *Minding mind*. Boston, MA: Shambhala.
Dumoulin, H. (2007). *Zen enlightenment*. Boston, MA: Shambhala.
Elliot, T. S. (1968). *Four quartets*. New York: Harcourt Brace & World.
Forman, R. (1999). *Mysticism, mind, consciousness*. Albany, NY: SUNY Press.
Guo Jun. (2013). *Returning home*. Retrieved June 20, 2015, from http://www.tricycle.com/web-exclusive/returning-home
Huxley, A. (2009). *The Perennial philosophy*. New York: Haperperennnial.
Ireland, J. (1997). *The Udana and the Itivuttaka Sutta*. Kandy Sri Lanka: Buddhist Publication Society.
Jayasaro, A. (2008). *Ajahn Chah's biography by Ajahn Jayasaro by the noble path*. Retrieved January 8, 2015, from https://www.youtube.com/watch#1-43
Kapleau, P. (1965). *The three pillars of Zen*. New York: Random House.
Kwong, J. (2003). *No beginning no end*. New York: Harmony Books.
Leighton, T. D. (2003). *Cultivating the empty field*. Boston, MA: Tuttle.
Leighton, T. D. (2015). *Just this is it*. Boston: Shambhala.
Loy, D. (1997). *Non-duality: A study in comparative philosophy*. Amherst, NY: Humanity Books.
Mitchell, S. (1993). *The enlightened mind*. New York: Harper Perennial.
Nananda, B. (1974). *The magic of mind*. Kandy, Sri Lanka: Buddhist Publication Society.
Pine, R. (2006). *Platform sutra*. Berkeley, CA: Counterpoint.
Roebuck, V. J. (2010). *The Dhamapada*. London: Penguin Books.
Sekida, K. (1975). *Zen training*. New York: Weatherhill.
Seung Sahn. (1976). *Dropping ashes on the Buddha*. New York: Grove Press.
Sharf, R. (1995). Buddhist modernism and the rhetoric of meditative experience. *Numen*, *42*(3), 228–283.
Sheng Yen. (1982). *Getting the Buddha mind*. Elmhurst, NY: Dharma Drum Publications.
Silaratano, B. (2009). *Mae Chi Kaew*. Udon Thani: Forest Dhamma Books.
Suzuki, D. T. (1978). *The Lankavatara sutra*. New York: Grove Press.
Waddell. (2011). *Wild Ivy: The Spiritual Autobiography of Zen Master Hakuin*, Boston, Shambhala. Kindle edition.

CHAPTER 6

Lives of the Awakened

Introduction

The knots that secure the net of meanings of the forest and mountain monks are the life stories of their spiritual adepts. These intersections are the narratives of the great accomplishers and their web-like connections with one another. Without these identities, the story of these movements unravels into random threads. Certainly, there is legitimate study focusing on the social and political history of both movements which often suggests a reality different from the idealized constructions of the lives of the awakened as told by Ch'an/Zen or the Kammatthana monks. Our interest is in their bio/autobiographies is not meant to deny historical or social investigations that seek the "real" story of Ch'an/Zen and the Kammatthana. Rather, it is recognition of the importance of how actors record and advertise their stories. Historically inaccurate constructions are social realities too. Meanings and imaginings matter.

Revivalist Buddhist movements such as Ch'an and the Thai Forest School are shaped by their soteriological project. The ultimate and proximate telos is liberation. The method of the project is pragmatic in its subordination of means to the ends. The methods entail self-cultivation and face-to-face interaction with an accomplished teacher. Each teacher's face is a story of struggle and attainment and behind the teacher are other faces reaching back into antiquity that tells a similar tale. The records of the Kammathana and Ch'an/Zen are the narratives of their teacher's lives.

A.R. Lopez, *Buddhist Revivalist Movements*,
DOI 10.1057/978-1-137-54086-7_6

HAGIOGRAPHY

These life stories can be termed hagiography. The type in its narrow and historical setting derives from the early Middle Ages in Europe and Christendom. Used in broader and less restrictive manner, the term can be applied to any laudatory biography or autobiography of the life of an exemplary figure of any social movement even secular groupings. In our age, Mao and Che are included along with medieval saints. Contemporary use of the term is often pejorative, suggesting an unrealistic work filled with self-serving, one sided praise. No such criticism is implied here. To see these works on the lives of the awakened as hagiography is meant to understand them as archetypical stories that allow for comparison and the isolation of general themes along with grasping their role in the world of Zen and the Forest monks.

Tiyavanich (1997, p. 11) disputes Tambiah's (1976) interpretation of Forest Tradition hagiographies as the Buddha's story cloaked in regional garb. She references her broader sample of biographies, many of which appear in Thai magazines, to bolster her case that Kammatthana literature is not hagiographic. In the examination below, a larger view of hagiography is taken that draws from cross-cultural studies of rites of passage and myths. The Buddha's biography is understood as a case of a larger category: the hero's quest. What generates the fascination of the Thai public is not the cloistered monk but the warrior-hero in the wilderness. The popularity of these tales which regularly appear in Thai newsstand publications testifies to the self-presentation of The Forest Tradition and the vital role of monk biography. Finally, the monks and mae chee not infrequently report recalling with emotional impact the life of the Buddha and events correlating with their own journey. Clearly, the Buddha (and other lineage teachers) is the deep template from which their lives are being lived and shared.

The narratives of the awakened merit attention because they perform a list of vital functions. An overview identifies the following:

1. Legitimacy—The narratives establish the spiritual credentials of their subject/protagonist and hence of their students and the community. Tales of awakening confirm that the sacred has been touch and embodied.
2. Guidance—Commonly, these life stories offer explicit guidance on how to practice. The life of the saint is a model for walking the path to liberation.

3. Inspiration—To traverse the road to freedom demands more than intellectual agreement. Motivation means the mobilization of value connected affect that the dramatic narratives arouse.
4. Marketing—Social movements exist in social milieus. Monastics depend on a lay world for material support and recruitment. The lives of the saints define the movement to outsiders and are an advertising vehicle soliciting support.
5. Social Cohesion—For religious orientations emphasizing other worldliness, a common identification is critical for social solidarity. The hagiography makes common and current deceased ancestors by bringing into the moment a history antecedent to the present teacher–student relationship. A shared lineage communicated through life narrative builds an imagined family.

Myth and Passage

The narratives of the awakened have recurrent themes which display the core values and worldview of the mountain and forest movements. Joseph Campbell in his seminal *The Hero with a Thousand Faces* finds the hero to be the archetypical protagonist in myths of quest and adventure across the globe (Campbell 1972). Interestingly, he declares the Buddha and his story as the quintessential hero narrative. Here, the elements of the problem or challenge, the quest, the triumph, and the return come together to form the primary plot of a world religion. The chapters are distinct yet unified by the heroic personage of the Awakened One. The story has been told and re-told across the globe from the Buddhist lands of Asia to Hollywood.

Campbell's analysis of the hero is pertinent in that the Buddha's spiritual adventure is the blueprint for the biographies of the awakened masters. The Buddha looms over the horizon casting his profile over the literary landscape of both Ch'an/Zen and the Thai forest monks. Much as the early Christian martyrs had their suffering framed as Christ's passion reenacted, so Buddhist monks envision themselves as walking in the footprints of the Buddha. With this core identification, the power of these hagiographic accounts and their ability to fulfill the functions of legitimacy, guidance, inspiration, advertising, and social cohesion is apparent. For all their solitary practice, the monks of the mountains and forests never walk alone. The path has been cleared by the Buddha and the illumined ancestors.

The anthropologist A. van Gennep (1960) offers an alternate yet ultimately compatible model by which to frame the life process of the Buddhist saint. Whereas Campbell's approach identifies the structure of the story highlighting its particular events, meanings, and ultimately its telos, van Gennep's study identifies formal phases and their sequences of ceremonies in relationship to the social group and its domain of meaning or nomos. Like Campbell, his study is comparative and cross-cultural. He terms the processes he examines rites of passage. It may be objected that the stories of the Ch'an and Forest monks are not marked by rites but are more the freeform quests that attract Campbell. The sequenced passage that van Gennep identifies, however, also effectively marks off the monk's journey. In place of rites of passage, we may simply refer to phases of passage.

In his worldwide and historical survey, van Gennep conceptualizes what he deems "characteristic patterns in the order of ceremonies" (1960, p. 10). They are constituted by three typical phases: rites of separation, rites of transition, and rites of incorporation. He also terms them pre-liminal, liminal, and post-liminal. In essence, the initial stage dubbed separation or pre-liminal consists of the removal of the subject or a group of subjects from the normative community. This phase signals that the subject is no longer a community member and that his former status has been vacated. The initial phase is followed by the liminal or transitional phase. Van Gennep's use of the term "liminal" is key as it signifies a threshold or edge where the actor is no longer of this world or at least any stable social identity. A transition or transformation takes place. The process is completed with incorporation back into the community (p. 11). Notably, van Gennep's conceptualization applies both at the level of the social group as a demographic, physical entity, and as a community of meaning. The model, like Campbell's, offers a framework for clarifying the life journey of the Buddhist adept.

Finally, before turning to the actual, individual accounts, an obvious contrast between Christian and Buddhist hagiography should be noted. Based on the different central narratives of Christianity and Buddhism, the primary archetype is different. The Christian saint is a martyr. The Buddhist saint is a heroic and victorious warrior. The martyr wins by losing; the hero wins by winning. Although the term hagiography is being applied in both cases, this distinction should be recalled. The functions of these laudatory accounts may be similar, but they are achieved through contrasting archetypes and storylines.

Sample Cases

In examining and then comparing the recorded lives of Ch'an/Zen and Forest masters, one is faced with a varied, rich, and challenging literature. Ch'an/Zen is represented by its 1500-year history spread over numerous dynasties and at least four modern states (China, Japan, Korea, and Vietnam) and starting in the last century worldwide, particularly in North America and Europe. Our inquiry will of necessity be selective. Biographies and autobiographical statements will typify different historical epochs and regions. Based on the availability of English language sources and the quality and quantity of translations and commentary, our regions of choice will be China and Japan. The historical spectrum will span what Fergurson has called the legendary period of the first four ancestors of Ch'an through the classical, middle, and late period in China. The Japanese representatives begin with Medieval Japan but will extend into the modern era of twentieth century teachers. North America and Europe offer several extensive biographical efforts on the transmitters of Ch'an/Zen to the West.

Even if not specifically cited, our review includes the more extensive and available life records of the following teachers: From China, the Legendary Ancestors, Bodhidharma, Huike, and Huineng; from post T'ang Dynasty Han Shan, Wu Wen, Hsueh Yen, Meng shan, and Kao Feng. The Japanese contribution features Dogen, Bankie, Bassui, and Hakuin. Obviously, the selections are not exhaustive, but they do encompass the more extensive accounts. Information on the lives of these teachers is also available from the works of contemporary scholars in addition to classical texts. While the former offer valuable checks and balances on the latter, it is the collections of the Ch'an/Zen sect that are most pertinent. Our interest is in how the movement portrays its ancestors to its adherents and the public.

In contrast to the wide ranging history and geography of Ch'an/Zen, the Thai Forest movement, the Kammatthana, offers a compact and well-circumscribed literature that one would expect of a movement born in the twentieth century and primarily located in one region, Northeast Thailand. The core teachers have all either produced autobiographies or been the subject of biographies. Like the pre-twentieth century Zen literature, these works are authored by monk disciples. Brief biographical sketches consisting of a few paragraphs of basic personal data on the lives of Thai ajahns are available on the web. Full-length works, some as formal biographies, and others as volumes of collected recollections, have been authored by long-term students.

The seminal biographical work of the Kammatthana is Maha Bua's *The Venerable Phra Acariya Mun Bhuridatta Thera Meditation Master* which has passed through several editions. This life story of Maha Bua's teacher and the initiator of the modern Thai Forest Movement projected the Kammatthana into the national consciousness and in its English rendition (1976) into a global arena. Maha Bua's relentless and unqualified praise and promotion of his Master took the Forest Movement out of the forest and into the city. Maha Bua has also contributed a biography of Ajahn Khao Analayo entitled *Venerable Ajaan Khao Analayo A True Spiritual Warrior* and translated into English (2006) by Ajahn Pannavaddho, a senior disciple of Maha Bua. Another close affiliate of Maha Bua's, Ajahn Dick Silaratano, has compiled from Thai sources and his own notes a biography of Mae Chee Kaew (2009), a female enlightened disciple of Maha Bua. While the Maha Bua sublineage is the major source of Thai Forest teacher portraits, there are other important contributors. Notable are the autobiographical pieces of Ajahn Lee and Ajahn Thate. Lee's life story was posthumously compiled by Thanissaro Bhikkhu, while Ajahn Thate's *Autobiography of a Forest Monk* (1993) was published when he was over 70 years in robes.

There are numerous sources on the lives of the two most prominent forest teachers, Ajahn Chah and Maha Bua. In print are several collections of reminiscences and expositions on their lives by close monk disciples. A series of YouTube lectures by Jayasaro Bhikkhu constitutes an extensive series of audio/video talks on the life of the perhaps the most influential forest teacher for the West, Ajahn Chah. Ajahn Amaro has authored what he entitles *An Introduction to the Life and Teaching of Ajahn Chah* (2012). Faithful to the title, Amaro's work places Chah in the context of the Forest Tradition, traces his personal life, and then gives an overview of his message and methods of instruction. Ajahn Jayasaro is also intending to produce a biography on Chah. Maha Bua's recent passing has also inspired a number of new accounts of his remarkable career. The following constitute the core Thai Forest personalities that are the subject of significant auto/biographical works: Ajahn Mun, Ajahn Khao Analayo, Ajahn Thate, Ajahn Lee, Ajahn Maha Bua, Ajahn Chah, Mae Chee Kaew, and Ajahn Liem.

In addition, there are numerous abbreviated pieces ranging from a paragraph to a page in length of other forest teachers. Many can be found either as brief introductions to their dhamma talks or online at accesstoinsight, forestsangha, forestmeditation, or at websites representing their wats. The most extensive reviews in English are to be found in Tiyavanich's *Forest Recollections* (1997) that focuses on the stories of ten forest teachers.

The Primary Narrative

A Tibetan lama once told that an enlightened layperson in his village recounted that his moment of awakening occurred when he met his guru who appeared to him as "just like the Buddha." A Thai graduate student researching the Forest Movement insisted that the power of the school derived from teachers who were "just like the Buddha." Both the forest and the mountain movements claim to preserve and revive the way of the Buddha. The Forest Movement has its Vinaya, code of discipline, and Zen has its wordless transmission via a raised flower. Although in slightly different ways, the Theravada touting its texts and monastic routine and Ch'an/Zen its mind transmission claim the Awakened One as their primordial guide. The Buddha's story is the primary hagiography. Yet in the initial centuries after the Buddha's parinibbana, his followers showed minimal interest in the founder's personal story. Buddha images were not fashioned and his biography not written. We have to wait until the second century C.E. before the poet Ashavagosha composes an epic on the Buddha. Three sutric texts and commentaries, all composed in the C.E., are the primary traditional sources. The *Lalitavistara Sutra* is preferred by the Tibetans, while the *Abhiniskramana Sutra* is the most complete of the early texts and has been translated several times into the Chinese. The *Nidanakatha* dates from the fifth century and was composed by the great Lankan scholar-monk Buddhaghosa.

The details vary but the overall storyline has a similar contour. Although relatively late in composition, obviously a need was eventually felt to tell the Buddha's story. Despite the intervening centuries, the importance of hagiography for the broader tradition took form. Most Asian school children and laypersons know the Buddha's life tale far more than dharma. The life of the Buddha can be framed by van Gennep's three movements: pre-liminal, liminal, and post-liminal. In Campbell's view, the traditional Buddhist narrative identifies a sequence of events toward, into, and back from the liminal.

The interpretive frameworks of Campbell and van Gennep converge at the concept/stage of liminality or threshold. This phase which is the climax of Campbell's hero-quest is also the pivotal passage in van Gennep's model of rites of passage. Prior to coming to an existential threshold and entering into a non-ordinary realm, the protagonist must separate from his ordinary life. This is the pre-liminal act that begins the movement toward the liminal. Turning to the Buddha story, which Campbell repeatedly cites as an example of a more universal tale, several sub-stages are noted: birth and prophecy, the call and dispassion, departure and the threshold of separation, and the road of trials (1972, p. 25ff.).

Forest and Mountain Tales

The Thai Forest Tradition positions itself as standing in front of the ahistorical mural of Theravada Buddhism. Its literature is the Pali Canon and what we know of persons is limited to the authors of commentaries and the leaders of expeditions to Sri Lanka and Southeast Asia. The Forest Tradition offers no compendium of masters transmitting the Dhamma from mouth to mouth. The symbolic context for the Kammatthana is the classical texts and not a line of teachers. There is the Canon and the monk standing before it. For instance, we know little of Mun's ancestors as individual personalities nor has the Kammatthana attempted to construct a fanciful lineage stringing together the contemporary monk with his forefathers. In the Theravada liturgy, there is no lineage chant like the one intoned in the Zen monasteries. Yet the Kammatthana do show a preference for the personal story not to be found in other Thai lineages. The non-personal style of the Theravada in general does not so much cancel the hagiography, as much as it points to the unique, innovative, and vital function of the biography in the Kammatthana's imagination. The non-personal static doctrine of Theravada becomes distinct personalities. With the emergence of Forest Movement's biographic literature, the landscape of the forest and the mountain touch in their preference for the personal life story.

How does Ch'an/Zen tell us about itself? The very titles in the Ch'an library are revealing. The *Gao Seng Zhuan* (Biographies of Eminent Monks) and the *Xu Gao Seng Zhuan* (The Continued Biographies of Eminent Monks), the *Chan Lin Seng Bao Zhuan* (Treasured Biographies of Chan Monks), the *Fo-tsu Li-tai T'ung-tsai* (Extensive Historic Record of the Buddha Ancestors), and finally the famed *Transmission of the Lamp* are the core collections and all are organized around the life and teaching of notable Ch'an persons. There are other sources primarily the records of individual teachers such as the *Zhenzhou Linchi Huizhao Chanshi Yulu* or the record of Linchi (Rinzai, Jp.). There are fascicles of Ch'an verse, *Trust in Mind* by Sengtsan, and collections of kung-an (koans), such as the *Wumen Guan*, the renowned Gateless Gate. While not constituting a true biography, still it is the personal anecdote that is the building block of Ch'an/Zen's imagined history.

The organizing principle of the Zen school's story is the individual teacher rather than information of its growth (and decline) as an institution, institution to institution relations, or key events in its collective history, or even a systematic elaboration of its thought. Instead, the unit of self-presentation is the person of the master. Strung together like so many beads, they constitute the history of Ch'an/Zen. Scholars are forced to break new ground to move beyond the singular teacher. Such efforts are for those on the outside trying to look in, but when Ch'an/Zen wants to talk about itself, its lexicon is the life stories of its masters. The forest school imagines its story as "just like the Buddha," but the Buddha in the forest of Isan. The Zen school, as we shall see, developed a distinct hagiographic template that while not abandoning the Buddha as a figure of inspiration, synthesized a new persona: the Zen master. The Buddha's story is sharpened with the tale of Bodhidharma to give a Zen edge to the figure. The forest movement positions itself alongside the Buddha, drawing its silhouette from the great arahant disciples. What the two schools maintain in common is the vital place of the teacher and his story.

Pre-Liminal: Birth, Prophecy, and Separation

The birth of the Buddha and the subsequent days of his early life have been imbued with all the miracles worthy of the founder of a great religion. He awaits birth in the Tushita Heaven and is introduced into our world via a white elephant impregnating his mother. The future Buddha emerges walking and talking and in Mahayana versions announces, "In heaven above and earth below I am the only One." Along with this breath-taking entrance, a prophecy is being made back at the court. The newborn will either be a world emperor or a world redeemer. While no accounts of birth and childhood of forest masters and Zen adepts dare rise to this level, they are not without heraldic signs.

Like Mun, his biographer, Maha Bua, comes with early signs of spiritual notoriety. While in the womb, his mother noticed that the fetus would be still for extended periods of time which concerned her about a possible intrauterine death. Then the future arahant would exhibit a burst of activity. This alternation between stillness and fierce action was later interpreted as indicators of Maha Bua's capacity for deep samadhi practice and vigorous teachings. His grandfather understood it as a sign of a boy with great determination. Like the Buddha, Maha Bua's birth was accompanied by prophecy. Upon seeing the placenta slung over the infant's shoulder, grandfather saw the alms bowl strap and predicted a monk's career. He also predicted a possible career path as a hunter or a thief. Maha Bua's kamma was the former (Foong Thim Leng 2011).

Maha Bua's student Mae Chee Kaew displayed signs of spiritual precociousness throughout her childhood. As a child, she communicated with devas, spiritual entities recognized in the Buddhist pantheon. By the age of 7, she began to recall past lives. Such extrasensory skills marked the young Tapai as a woman of great spiritual potential (1960, p. 26, p. 29). These faculties would continue to operate during her lifetime at times energizing her practice but at other times deflecting her from the path to Nibbana. Han Shan's mother had a prophetic dream of Kwan Yin as a result of which she became pregnant (Chang 1959, p. 118). Huike, Bodhidharma's heir, was presaged by a strange light in the sky by which his mother became pregnant. Hongren's rejecting mother left him lying in his embryonic fluid. Upon her return, he had miraculously cleaned himself (Ferguson 2000, p. 21). The birth/childhood indicators establish the promise and give direction to the story of the teacher to be. While the Buddha is portrayed as a maestro at birth, the masters of the mountains and forests are promising child prodigies.

The Call and Dispassion

Like the chick breaking the shell, the encasement in ordinary life, often village life, needs to be cracked for the journey to commence. In the Buddha's story, it is the walls of the palace which require opening, and the famous Four Signs, old age, sickness, death, and renunciation, make vivid the inadequacy of ordinary life and beckons to the beyond. Typically, the stories of the masters tell of discontent, sometimes shock that propel the protagonists and loosen their entanglement with the world. Dispassion is followed by a call to the enunciate life.

Ajahn Lee's autobiography touches many of the key markers of the Buddha's formative years and early adulthood minus any miraculous birth omens. Cardinal to Buddhist dispassion has always been the insight into impermanence. The first three of the Four Signs the Buddha encounters on his sojourns outside the palace may be seen as indicators of the impermanence of life. Birth and death have been viewed with trepidation and been a motivation in Buddhist circles. Lee recounts his near pathological revulsion at both birth and death. He describes his childhood encounter with a village woman giving birth which in traditional manner was accomplished by holding onto a rope. Her screams and moans filled the young boy with "fear and disgust" (1991, p. 3). Death was equally terrifying for Lee. His mother passed away when he was 11; he refused to attend the funeral and later would scrupulously avoid any food or implements from a funeral ceremony. As a young monk, he dreaded having to chant at funerals (p. 5). In an unmistakable replication of the Buddha's Four Signs of old age, illness, and death and the renunciate, Lee tells of four events which warned him of the lay life. The incidents while not remarkable in themselves were taken as omens. "All these event I took as warnings" of worldly thought of disrobing. The Buddha's disenchantment with the world and the awakening to the quest are replayed by Ajahn Lee (1991, pp. 29–31).

In the life stories of Zen teachers, death is often the catalyst and at an early age. Dogen's quest was also put into motion with a funereal confrontation with death. Watching the smoke rise from a stick of incense, the young boy had a deep insight into the impermanence of all things. The door that had opened to reveal the transience of life was also the door that closed on the world of family and career. Dogen was placed on the path of practice and eventual awakening. Bassui asks how his deceased father will eat the food offerings at his memorial service. Dissatisfied with the explanation, his quest begins (Braverman 2013, p. xiii).

Along with his fear of birth and death, Lee also had a deep and understandable ambivalence about domestic life. He vowed to never marry then relented but soon found marriage and family disappointing and returned to the robes. Lee aversion to three marks of existence in his case, birth, death, and marriage, seemed to have generalized into distaste for village life. "I have to leave this village," he announced to his father (1991, p. 7).

Like the Buddha, there was also a draw to an alternative. The Fourth Sign in the Buddha's story showed the way of renunciation. Ajahn Lee tells of attending a dhamma talk. On the sermon seat was a monk he had never seen before. "I was really taken by the way he spoke," he recalls (1991, p. 6). Inquiring about the monk, he was told that he was a disciple of Ajahn Mun, the meditation master. Lee visited the monk's encampment and was impressed with his way of life. Then, he set out on foot to find Ajahn Mun. Needless to say, the meeting was momentous for the young Lee and his journey had begun in earnest.

Ajahn Khao Analayo, whose biography was penned by Maha Bua, experienced disillusionment by way of a secondary meaning of the word *annica*, unreliability, namely his wife's infidelity. Catching the couple in flagrante delicto with machete in hand, Analayo reflected on what he was about to do. Dhamma filled his heart and the awareness of the law of kamma, the moral teachings of the Buddha, and the futility of lay life all flooded in. "I worked hard but often we had to go without." He observed about his lay existence "Going beyond the world to attain Nibbana following Lord Buddha and his Arahant disciples was the only course I was willing to contemplate" (2006, p. 22). No doubt to the relief of Analayo's wife and her lover, the future arahant broke with the world and began his journey to Nibbana.

Ajahn Thate who had direct contact with Ajahn Mun, but studied primarily with Ajahn Singh, a senior monk of Mun's, gives a more naturalistic account of his life in his *Autobiography of a Forest Monk* (1993). Supernatural entities that abound in Mun's accounts as "objective" forces are largely confined to Thate's dream life. Thate's disillusionment comes via a nighttime reflection on the hardships of the farmer's life. In a remarkably mature and sustained consideration for a young boy, he reviews in detail the annual work cycle of the Thai peasant—a life he intimately knew. The First Noble Truth, dukkha, and the cycles of dukkha, birth/death, and samsara become vividly immediate. "The future leads on to continuing doing," he concludes. "I clearly perceived all the suffering involved in being born into this world" (1993, p. 42). Thate story also contains a prophecy echoing the Buddha. The prediction comes from Ajahn Mun himself. Thate is discrete and does not reveal the content other than it was a forecast of a successful dhamma career and his own embarrassment at its pronouncement (1993, p. 70). Although both the event of dispassion and the prophecy are presented in Thate's autobiography in everyday naturalistic tone, they replicate key events of the Buddha's life. Without any exaggerated claims, which Thate avoids throughout his autobiography, his self-presentation and self-understanding place him in the footprints of the Awakened One.

The early life of Mae Chee Kaew, an acknowledged female arahant, repeats the common theme of untimely death. The Buddha's mother Maya Devi dies from childbirth, Ajahn Lee loses his mother, and so does Mae Chee Kaew. At the age of five, her mother dies collapsing little Tapai's world (Silaratano 2009, p. 27). Her father's remarriage produces a half-brother, who also unexpectedly dies (2009, p. 31). Death is proximate and unpredictable. Nevertheless, there is something more than mere death. By the age of seven, Mae Chee Kaew is recalling past lives and being recognized as a precocious, if unstable, young girl by the likes of Ajahn Sao and Mun.

Mun experiences what his biographer calls a Great *Samvega* or profound disillusionment with the world. He sees into the folly of human beings including his own that exposes the futility of the worldly life and by contrast the necessity and attraction of the Buddha Way. His mind entered a state of detachment while in "a state of sheer 'emptiness'" (Maha Bua 1976, p. 27). Just subsequent to his ordination, Mun has a dream of a white horse which he interprets as an indication of Nibbana's availability and the correctness of his chosen path.

While these events of disillusionment and prophecy lack the clear mythology of the Buddha's narrative, they convey an identical learning. The worldly life is without lasting reward. Several of the stories have the advantage of earthy, everyday occurrences, the agony of childbirth, the suffering of sickness, loss through death, the grinding labor of farm work, and the heartbreak of faithlessness. Any lay Thai can relate to them. They imitate the Buddha, but they also couch the message in a vernacular that translates into the everyday lives of villagers and townsfolk. The surface of the story is local and common, but the deep structure recapitulates not just the Buddha but an archetypical quest/hero story.

Departure and Separation

The classic rite of departure for the Buddhist is ordination which re-enacts the Buddha's cutting of his hair and surrender of his material possessions. Van Gennep notes that separation often entails the assignment of a new name and literal physical separation from the community. Receiving a dharma/dhamma name is of course an aspect of ordination for both Theravada and Mahayana along with the more visible act of shaving the head. In some cases, novices may be sent far from home but even when not they are denied contact with family and friends. The ordination can be seen as a complete rite of passage in itself. Interestingly, the Thai *buat* or ordination spans three days making a tripartite process that matches van Gennep's model. The three days conclude with the novitiate being presented to the community, who in turn offers alms, thus confirming his new status.

Such ritual is not confined to Zen or the Kammatthana. What the forest movement and Zen add, like the Buddha's plunge into the jungle of India, is a further separation that brings the monk to a more profound inner quest. Each of these movements is identified with a natural environment in which they practiced and established monasteries. The mountain and the forest are more than plants and rocks. They are the symbolically charged realms where the quest is under taken and consummated. They stand in contrast to the world of the city and the village. This is the unmarked terrain where the challenges and trials must be faced if one is to join the company of the Victorious Ones.

The auto/biographies place little emphasis on entering the monkhood, rather it is joining the dhutanga or thudong life style which constitutes the critical transition. This is in keeping with their distinctive project and the way of its realization. One of the more dramatic entrances to the journey is Ajahn Liem's charnel grounds practice. As a novice not yet eligible for full ordination, he was determined to practice in the local cemetery where corpses were often left unburied and at night was a home to wild dogs. "All my hair would stand on end, and I would feel like screaming out. But I couldn't," he recalls. "I'd think. If I have to die, I'll die" (Chah 2005, p. 80). Having confronted death, he is no longer a worldling, although still a distance from his destiny.

The home leaving of the Buddha becomes a story within the stories of the forest monks and Mae Chee Kaew. After her decision to leave home, Mae Chee Kaew falls into a pensive mood in which she reflects on the Buddha's parting from his family. She self-consciously identifies with the Buddha crossing the threshold just as her biography documents the dramatic break with her family. Mae Chee Kaew's teacher, Maha Bua (Foong Thim Leng 2011), speaks of his being brought to tears as a young monk upon reading the Buddha's story. Such accounts, conveyed within their own biographical material, strongly suggest the underlying template of their lives as seen by the protagonists themselves.

Road of Trials

The journey to awakening is not for the weak of heart. As we have already seen in examining path and practice in the two schools, their shared controlling metaphor is the warrior. There is no warrior story without an enemy even if only of circumstance. If the protagonist is truly a hero, then that status needs to be forged on the road of trials. Before our hero can enter the liminal proper, he needs to pass the test of determination and skill.

The Buddha's trials are largely self-inflicted. He takes on ascetic practices, starving himself till his abdomen touches his backbone. He controls his breath inducing knocking sounds in his skull. The Awakened

One-to-be speaks eloquently of the terror of the forest and its enveloping aloneness. The sounds of jungle beasts and the rustling of leaves all portend threats both mental and corporeal. Even the Buddha's apprenticeships with the two forest gurus, which may at the surface seem supportive, are also their own trials. In triumphant mastery of their teachings, he dismisses the temptation of becoming their spiritual heirs for the challenge of further trials.

The Thai Forest Movement prides itself on walking the path "just like the Buddha." Forest monk's informants were very aware of the homologous fit between the Buddha's way of practice and its privations and those of their teachers. Later in life, Maha Bua vehemently chastises others for their lax practice invoking his own endured privations (2005a, p. 89). The trials and tribulations of the way link teachers and students in meanings that are not merely cerebral but visceral. Maha Bua weeps when recounting the sufferings of Ajahn Mun. Like war comrades, there is a bond that transcends social background and personality.

Facing down wild elephants and tigers is a recurrent test on the path of the forest dwellers. Such stories are no doubt genuine in early twentieth century Thailand. Even today monk informants have told me of observing tigers during their morning meal. On a forest walk I enquired if there were still tigers about. "People who camp here don't come back with their dogs," I was told. Encounters with elephants seem to be the favorite anecdote, perhaps because they replicate an incident from the Buddha's life when he pacified an elephant set upon him by his cousin Devadatta. Such tales have a heart stopping drama not matched by enduring ants in one's robe.

The life tales of both Lee and Khao Analayo contain such pieces. An elephant in rut had been terrorizing the local village where Lee was camping. When the beast approached his encampment, the villagers urged him to flee but they dared not help. Relying on metta meditation, as did the Buddha, Lee quieted the beast, who with a flap of his ears departed (Lee 1991, pp. 18–19). Analayo has a similar story. Hastily setting up a circle of candles he began walking meditation, calling on the power of the Buddha by reciting *buddho*. The elephant halted his charge and stood looking at the monk for perhaps an hour before turning away to forage for food. Such events fostered deep faith in the protagonists but also in the hearts of the villagers. "The heart bows down putting faith in the Dhamma," Khao Analayo's biographer concludes (Maha Bua 2006, pp. 38–39).

Pachyderms are the animal of choice on the Road of Trials, but it was possibly that smaller creatures, insects, posed the greater threat. In a reminiscence, a former monk devotes several pages to the varieties of ants, each more ferocious than the former, encountered by the forest monk. Elephants may crush the forest monk, but ants regularly turned walking meditation into hopping meditation, as the practitioner needed to step over columns of militant soldier ants. Far more deadly but less grand was the trial by disease especially malaria. Probably more monks succumbed to illness than to their animal adversaries (Breiter 2004, loc. 680). That elephants and tigers are the threats mentioned in forest monk life stories suggests that drama and not banal realism is the guiding principle for the memoirs.

With the triumphs on the Road of Trials, the protagonist is worthy of entering into the liminal, the realm of sacred power. He/she has more than survived; they have qualified. The qualities of the hero have been developed and displayed, and the right to pursue the ultimate has been confirmed. One thinks of the Buddha after having endured the privations of ascetic practice, being challenged by Mara on his audacity to pursue final release. The great determination of the quest is supported by the confidence of both an inner and outer power. The heart is pure and focused, and as Analayo tells us quoting the Dhammapada, "The Dhamma guards those who practice the way" (Maha Bua 2006, p. 38).

In the modern Ch'an tradition, we have several contemporary accounts. Hsu Yun, arguably the outstanding master of the first half of the twentieth century, tells the story of his quest in his autobiography, *Empty Cloud*. In his 56th year, he headed out to Gao-min monastery to attend a retreat. Unable to afford the cost of a ferry, he wandered along the banks of a rapidly rising river and slipped and fell in. For an entire day and night, he bobbed in the waters in a near death state until fished out by a local. By his own account of bleeding from all his orifices including his genitals and anus, he manages to arrive at the monastery to face his final test. One is challenge to think of a more harrowing trial in an effort to literary gain the other shore (1988, p. 65).

Sheng Yen, a Taiwanese-based monk with international connections, also offers his autobiography, *Footprints in the Snow* (2008), and leaves behind shorter hagiographical works by his devotees. Both these Ch'an monks suffer the poverty of the ordained and the shortages that came with China's twentieth century upheavals. Sheng Yen escaped to Taiwan where conditions were initially not easy, but Hsu Yun remained in mainland China and had to endure persecution by the communist authorities. Stories of his physical abuse by communist soldiers suggest the hagiographies of Christian saints, but true to Buddhism his final stature rests not on his suffering but on his transcendence of suffering (Porter 2009, p. 327ff.). Sheng Yen's privations extend to his pilgrimage to the West, although this account has been disputed as an exaggeration designed to ape traditional Ch'an Buddhist hagiographies (Lachs 2011). True or not, it underlines material deprivation as stylized and sometimes very real trial in Ch'an/Zen life stories.

Encounters with wild beasts are not a Thai monopoly. One of the more dramatic confrontations with wild life is found in Hakuin's autobiography. He tells of his teacher Shoju's meeting up with wolves. Gathering in the cemetery where Shoju sat, wolves in great numbers tested the monk. They charged him only to leap over him at the last moment. Sniffing and butting him with their snouts, he could feel their warm breath. But with true Zen resolve, "Shoju never once flinched or wavered." Instead, he reports a secrete joy over his courage and strength. He had sat through his seventh night of meditation (Waddell 2011, loc. 1775).

Bankie endures 14 years of extreme privation that compromise his health and culminate in tuberculosis. His buttocks became swollen and bloody from sitting on bare rock. "Where ever I spat, gouts of bloody sputum as big as thumb heads appeared." Only when his illness reached a decisive stage did Bankie breakthrough to awakening (Waddell 2000, p. 10). Bankie's suffering recalls the Buddha's extreme ascetic phase. The liminal can only be entered by passing through trials even if self-imposed.

Although the Buddha rejected what he saw as the extremes of asceticism, the Road of Trials can be understood as a process of purification that prepares the protagonist for the great battle ahead. Campbell notes that the trials are not gratuitous problems, but that the hero is being cleansed to receive the ultimate attainment. In Theravada, far more so than in the Mahayana Ch'an/Zen world, purification, the expunging of defilements is a prevalent theme. The tests of the forest make one pure and allow approach to the threshold. Also we should remember that the dhutanga practices of the forest monks were the Buddha's acquiescence to demands for a more ascetic practice. For Ch'an/Zen, the trials affirm central values: courage and discipline. The Zen master is a Zen master because he acts like a Zen master. For both schools, the master to be is first a warrior.

The Liminal

The liminal (van Gennep) or the threshold is the realm beyond the known that is a "zone of magnified power" (Campbell 1972, p. 155). Here is darkness and danger but also liberation. In another framework, this is sacred power as opposed to the profane world (Durkheim 2008, p. 33). In Buddhist thought, this corresponds roughly to the Unconditioned, *asankhata*, contrasted with the conditioned world of causes and conditions. The Road of Trials leads here but the way is by no means clear, and additional dramas are still to be enacted. Supernatural forces may bar the approach. In addition, the dhamma hero may enter a dark night devoid of a guiding teacher and have to be able to find their own axis mundi for the final assault on peak of enlightenment.

At the threshold of the liminal is the supreme and final barrier: the Guardian. Although encountering and mastering the guardians of the liminal could be seen as one more additional trial on the road, it can also be understood as decisive. Campbell notes that the threshold is typically guarded by a super-human force who bars the gate to any presumptuous enough to attempt entry. In the classic Buddha story, Siddhartha's right to sit beneath the Bodhi Tree in quest of ultimate release is disputed by Mara. With his famous earth touching gesture, the Buddha-to-be calls upon the earth to be his witness thereby clearing the way to enlightenment. With the earth touching mudra, he crosses the threshold and enters the liminal. The decisive moment is at hand.

Ajahn Mun faces a guardian before securing the status of non-returner, the stage that precedes full arahantship. The encounter occurs in Sarika Cave located in what is now the national park of Khao Yai, Big Mountain. Mun finds that he has company in his hermitage: a giant who is uninterested in sharing his home with a dhutanga monk. Mun had been forewarned by the locals, but the dhamma warrior was undeterred. Like monks who had previously stayed in the cave, fallen ill, and died, Mun also suffered a health collapse that herbal medicine fails to reverse. Deciding to invoke the power of his meditation and Dhamma, he heals himself by entering the "luminous" mind which detaches from the body only to encounter a 30 foot black giant whose magic had been the source of his torment. Mun responds to his threats by invoking the purity of his cause and the power of Dhamma and the danger of kamma. Chastised, the giant transforms into a humble Buddhist layman. Like the Buddha beneath the Bodhi Tree, Mun calls upon transcendent forces and his right intentions to establish the legitimacy of his questas he sits under the Inner Bodhi Tree as his biographer describes it. (Maha Bua 1995a, p. 236).

The guardians of Ch'an/Zen also assume a non-human form. At time, they appear as doors or gates. Masters slam the door on the candidate, refuse to open doors, or throw applicant disciples out the door. The door itself can be the instrument of awakening as in the famous case of Yumen who awakened upon having his leg caught in the door by the master (Dumoulin 2005a, p. 231). The most prominent collection of Zen koans is the Gateless Gate, suggesting that even if there is no physical barrier, one still must pass through. Particularly unmanageable koans or phases of practice are called barriers of "silver mountains" or "iron walls" (Chang 1959, p. 148). Yet the ultimate guardian can be quite human: the Zen master. The master is the barrier who challenges the supplicant with verbal conundrums and, at times, physical assaults. The way to enlightenment is barred. Huike stands in the snow before Bodhidharma, young monks sit outside the temple, and even western lay students must face the master in doukusan.

The Way of Aloneness

The Buddha is the self-arisen and self-taught unlike his disciples who are hearers of the message. The final stage of his quest is without a guide or directive teaching. He must find his own way, albeit re-discovering the overgrown path of past Buddhas. No teacher in either school explicitly claims a status of self-arisen/self-taught. Yet a recurrent theme is that the final phase of the journey is without a teacher. No matter how important the teacher/student bond, which as we have seen, shapes the Forest Movement and Ch'an/Zen, in the end, one is alone. The difficulty of this passage is noted repeatedly by biographers and autobiographers in both schools. The practice/liberation traditions share the drama of the solitary, existential moment.

Ajahn Mun, for example, is reported to have searched for teachers, but no primary kalyanamitra is ever identified. Ajahn Sao is his senior but functions more as a stalwart dhamma brother. In his biography, Mun appears more as a solitary vanguard of a revitalization movement than as a student of established mentors guiding him down a well-worn path. Despite Mun's attainment, his students also face this phase of teacher-less aloneness. Mun attributes much of his suffering to the lack of a competent teacher (Maha Bua 1976, p. 16). He tells us "dangers and delays" are caused by the absence of a genuine guide (p. 11). Without any counsel, he was forced to move carefully on his ascent toward the peak of Nibbana (p. 100). Maha Bua, Mun's disciple-biographer, faces this plight as he only launches his assault on the heights of Nibbana after his master's death.

Despite the fealty to the teachers and the admonition to trust in the kruba ajahns, the recurrent theme of the teacher-less trial may reflect the truth of any authentic quest, but it also mirrors the path of the primal charismatic, Lord Buddha. In the final stage, the Buddha-to-be must find his own way. Rejecting the extreme of asceticism, he is abandoned by his companions and left with only his own intuitive wisdom as a compass. Maha Bua tells of his confusions and self-deceptions when navigating the upper reaches of the mind. In the final moment, the discipline and the teacher can only prepare the student for a lonely climb. As already mentioned, Mun attributes the extremes of his suffering to his teacher-less condition (Maha Bua 1987, p. 100).

Much of the Ch'an/Zen narrative is the search for the teacher and in some cases the post-enlightenment search for verification from a teacher. Seemingly, the more profound the quest the more stringent the standards and the more difficult it becomes to find a qualified teacher. Dogen exhausts Japan and leaves for China. Bankie is repeatedly disappointed even turning his back without a word on a Chinese master disembarking at the port. From afar he could see, he was not a person of the Way (Waddell 2000, p. 9). He would practice alone only resuming his search after his awakening. The quest for the teacher can be seen as part of the trials, but it also sets the stage for the drama of the solitary hero.

The Immovable Spot and the Great Battle

At the commencement of the Great Battle against ignorance and attachment, the Buddha takes his unmovable spot beneath the Bodhi Tree. Campbell designates this gesture as archetypical in the hero's journey. For Campbell, the Immovable Spot is the Navel of the Universe, the Axis Mundi, where the hero will make his stand or in the case of the Buddha take his seat. In various Buddhist traditions, this is known as the Vajra or Diamond-cutter throne. In the Vajrayana tradition, this aspect of the enlightened mind is personified as Akshobya Buddha, literally the Unshakeable One. The humble cave, earth beneath a forest tree, a cliff overhang, or even a meditation hall may lack the mythological grandeur, but they assume the same functional role as the Buddha's throne. The stories of the masters consistently stress the resolve and determination to sit through the coming storm of the forces of delusion and temptation "just like the Buddha."

Ajahns Mun, Analayo, Maha Bua, and Mae Chee Kaew are all portrayed as steeling themselves for a decisive battle with a resolve that itself is the immovable spot. (That each of these teachers is a member of the same sublineage is a matter we shall return to.) At times, the "spot" is a literal location, but what is essential is the determination to breakthrough to the Unconditioned even at the cost of their life. The description of Ajahn Mun's fight "to the finish" as his biographer terms it is the classic crescendo to the hero's quest or a contemporary action film. Nothing would be allowed "to penetrate the defense line of mindfulness-and wisdom" (Maha Bua 1976, p. 118). Even metta, loving kindness, is expunged from his mind, the dhutanga warrior monk launches his final assault. In his victory, all the worlds "seemed to tremble with awe and wonder" (p. 124).

Maha Bua employs similar language in his biography of Ajahn Khao Analayo (2006). To practice mindfulness is to be "locked in combat" with the kilesas, defilements. He quotes Analayo as referring to this stage of practice as "hand to hand combat." His biographer attributes his triumph to his determination and even aggression (p. 208). Ajahn Thate description conforms to Campbell's Unmovable spot and Great Battle motif. "I sat up and established mindfulness, settling the mind in stillness on a single object and ready to sacrifice my life" (Thate 1993, p. 170). Then in imitation of the Buddha, he confronts the temptations of Mara, the Evil One. In his case, Mara sends only one daughter in the disguise of a white robed Mae Chee. Thate also offers a more psychological analysis of his adversary as *anusaya-kilesa*, latent defilements buried deep in the mind. Nonetheless, this is "the field of battle where one can fight for victory" (p. 174).

Zen abounds in climactic stories of unshakeable resolve. Bankie, alone in his mountain hermitage, reaches the end of his rope. Suffering from monk's disease, tuberculosis, unable to eat solid food, he prepares to face his death sitting in meditation and facing the wall of his mind. With what might have been his final breath, he expels a blood clot. As the sputum slides down the wall, his mind breaks open exposing the always present Unborn Buddha Mind. He spent his remaining years teaching vast numbers, especially laypersons and women, that accessing the Unborn Mind need not be as difficult as he made it. Yet his story communicates the resolve of the Unmovable Spot and the Great Battle to the death against the adversary of ignorance. Throwing away his medicine spoon, he resolves to attain the Ultimate or die trying (Waddell 2000, p. 10).

Bankie is preceded by the monks of China. Hsueh Yen describes his push toward the ultimate. "From dawn to sunset I never left room or court." He looks straight ahead fixing his "eyes on a spot three feet ahead."

With a sudden turning inward of the mind, all "became clear, serene, and limpid." Meng Shan challenges himself with a refusal to sleep. First he lies on a pillow, then his arm, then sits up for days. The struggle culminates in a "mind like the clear sky of autumn or like pure snow filling a silver cup" (Chang 1959, p. 151). The great battle is found not only in past centuries, but also in contemporary narratives. David Chadwick's (2000, p. 41) appreciative biography of Shunryu Suzuki claims that his awakening occurred when the monk while attempting to steal melons from the monastery cellar was forced to stand stock still for hours when a metal hook caught in his eye. Suzuki's eye was saved and his wisdom eye opened making this one of the oddest trials on record.

Post-Liminal: Return, Integration, and Death

All good stories have an ending and there is none better than a return. For van Gennep, this is the stage of integration back into the community. For Campbell, the cycle of the hero's quest is closed; the circle is completed. The Buddha returns for his 45-year ministry. The records of the Zen masters and the Forest ajahns continue with post-enlightenment episodes of their selfless service. All lives end, even that of the Buddha. Some accounts of the forest and mountain teachers' dramatic return and final moments are replications of a primal template. Having attained the ultimate victory, Mun returns to the company of his dhamma brother Ajahn Sao. The meeting recalls the Buddha recently awakened seeking out his former companions to declare his victory. A direct parallel with the world of the Buddha is being proclaimed. Mun's narrative is the Buddha's story reenacted. The practice/realization of the forest monk is an *imitacion* Buddha. His hagiography is retelling of the Buddha's narrative.

A version of Dogen's return to Japan that I heard several times in Zen circles has him responding to the questions about what he learned and what he brought back. His answer is the classic radically concrete consciousness of Zen. "The eyes are horizontal, the nose is vertical." What he brings back are empty hands and a tender heart. Dogen's return is not simply a re-integration but a revivalist declaration. The template is the primal Ch'an scene of Bodhidharma and Emperor Wu. Dogen's words are as challenging and enigmatic as his bearded ancestor's. The tale may be apocryphal, but it is how Zen chooses to remember and present itself.

Typically, post-awakening story involves the accumulation of a multitude of followers and the establishment of monastic training centers which confirms the master's accomplishment. Hakuin, for instance, is presented as generating a spontaneous religious center. Hakuin's story comes full cycle as he returns to die in his home village which under his influence has become a major center for Zen training (Waddell 1999, loc. 482). Bodhidharma and his successors attract swarms of would be disciples (Ferguson 2012, p. 328). People of all ranks swell Bankie's retreats (Waddell 2000, p. 21).

The death of the hero closes the narrative, and Zen has made this event a hallmark of its hagiographies. Perhaps receiving more attention than awakening, the death of the master is the climax that synthesizes Zen's message about itself. The Zen adept is known by how he dies. Both the when and the how of the death are under the control or at least conscious cooperation of the teacher. Sitting upright or giving a great shout, Zen masters pass on embodying the values of power, discipline, and directness. The stylized pattern of Zen death is already present in the in legendary period of Bodhidharma and the generations that followed. Daoxin, the Fourth Ancestor, delivers a pith instruction reminiscent of the Buddha final words. "All myriad dharmas of the world are to be dropped away." He commands his disciples to understand, protect, and carry into the future this truth. Then in the classic Zen departure mode, he sits upright and dies. The Great Sixth Ancestor, Huineng, imitates the pattern (Ferguson 2000, p. 41). He delivers final advice and then exits in perfect composure. This scene was to be replicated for the next 1000 years of Zen. The scenario is so critical to Zen's self-presentation that when a contemporary master died drunk in a bathtub, the information was suppressed and replaced with a more Zen-like exit (Heine 2010, p. 247).

Heine and Wright (2010) in their study of Zen teachers conclude that "the accounts of the death of Zen masters bear remarkable similarity." They further conclude that these stories are saturated with legend and represent a refashioning of the history to conform to an emerging template. "All those stories," they find, "begin, proceed, and end in much the same way" (p. 248). Of note is that the Zen mythos is less an imitation of the historical Buddha and more a composite constituting a Zen archetype. The legends of Bodhidharma and the T'ang Dynasty greats reshape the Buddha's story as Ch'an/Zen's working model.

Mun death scene, on the other hand, appears to be carefully orchestrated with an eye on the Buddha's passing. Like the Awakened One, Mun is placed on his right side, the lion pose, but it is difficult to maintain. Finally, a position is settled on that is half on the side on the back.

Maha Bua records Mun as giving a discourse on the Dhamma's essential points and urging all to practice vigorously (Maha Bua 1976, p. 281). The language repeats the Buddha's messages in the *Mahaparinibbana Sutra*. Mun enters Nibbana "just like the Buddha."

Anti-Saints, Rascals, and Hags

No review of the literature of the lives of the awakened would be complete without acknowledging a character that sets apart the Zen sect from its forest cousins, the anti-saint. Zen records avidly commemorate the doings of rascals in and out of robes. This genre gives the world the Buddhist bad boy who's clowning conveys a freedom and unconventionality not to be found in institutional Ch'an/Zen. A visit to a local Zen or Ch'an monastery is unlikely to turn up a modern day Ryokan, Ikkuyu, Layman P'ang, or China's beloved mad monk Ji Gong (Shaw 2014). Nevertheless, their stories are part of a preserved, unofficial record of the school. These popular tales are hardly conventional hagiography and certainly do not imitate the Buddha's life, and yet they are biographies that expand the role prescriptions of the monk/practitioner. They proved relief from the more remote and wisdom laden lives of the great masters.

A cautionary is in order. The anti-saint is not necessarily seen as enlightened, but yet they carry an essential spirit of the way. Forest masters also are unconventional, as noted in their teaching tactics and gambits, but there is no literary genre (yet) that presents them primarily in this light. The radical tactics of the forest teachers remain within the markers of the precepts. Zen rascals indulge in sex, drink, play cards, and commit petty theft. Other than murder, they leave no precept unbroken. They communicate instead a freedom from convention which is the persistent subtext of Ch'an/Zen. While expanding conceptions of the holy life, they also create a new version of the standard hagiography and give expression to Bodhidharma's teaching of "nothing holy."

In contrast, the Kammatthana teachers are held up as scrupulous adherents of the Vinaya. We have encountered this variance between the forest and mountain repeatedly. The mountain monks react against a perceived staid Imperial Buddhist establishment. The forest monks react against a lax, superstitious world of village monks. Ch'an/Zen has a tendency to view rules as, if not hindrances, then at least secondary to the soteriological project, while the Kammatthana perceive rules as an essential framework for liberation. At the level of everyday temple life, rules are adhered to by both schools, but at the level of ideology and literary presentation, they display different leanings.

Along with the Zen rascal, an equally prominent figure that is absent in Thai forest narratives is the wise woman. In the Thai Theravada world women, ordained or lay, do not challenge the monk especially on questions of Dhamma. A woman may be a pure disciple to be respected or a temptation to be vanquished, but she is not a messenger of wisdom who rattles the male world of scholarship and status. Undoubtedly, Ch'an/ Zen stories are male dominated, but female actors, however marginal, are frequently the catalyst for the monk's journey.

The psychologist, Carl Jung, suggested that non-rational wisdom presents itself through a female form often appearing as the Old Hag who trips up the male identified ego. The Ch'an story of the monk carrying his voluminous writings on the Diamond Sutra who is unable to explain a pivotal phrase when confronted by an old woman on the road is classic. Burning his notes, he begins his quest for genuine knowledge. The nun Shi Chi walks into the temple carrying a basket on her head. If the monk will say a sentence of Ch'an, she will take down the basket. The monk is speechless. He invites her to stay. If he can say something, she will stay. Once again, the monk is flummoxed. Wisdom is not a one-night stand. If he wishes to know, he must commit to the quest for authentic knowing. The monk, later to be Master Chu Ti, recognizes that even though he is a man, he is ignorant of living Dharma (Luk 1970, p. 134). Perhaps the most charming expressions of Ch'an/Zen through the feminine are in the collection of anecdotes known as the stories of Layman P'ang. While male centered tales the Layman is often outdone by his equally enlightened wife and daughter. Daughter P'ang upon hearing her father's muttering that the dharma is "difficult, difficult" and her mothers rejoinder of "easy, easy" tops them both with her "not easy not difficult." Hearing of her father's intention to pass away, she climbs into his seat and beats him to it! (Ferguson, 2000, p. 96).

What is critical in these narratives is that the woman is not simply acknowledged in her own domain, the nunnery and the kitchen, but enters the male world to ignite the soteriological quest. The Thai Forest stories have no equivalent. The forest monk does approximate the Ch'an/ Zen rascal in his unconventional behavior from Ubaallii's inside out monk bag to Chah's barking like dog. Still the preferred forest master is the wise and disciplined monk. The anti-saint is not one of the stock characters in the forest narrative. This variance in the biographical tales of the two schools speaks to the more pronounced antinomian strain in Ch'an/Zen. The male-dominated world is not overthrown, but it is tweaked at times. Monastic precepts are observed, but the audience is reminded that playing beyond the bounds does not disqualify enlightenment but in fact may confirm it.

Critical Voices

Hagiography is not without its critics, even emanating from within the traditions. Questions raised about the veracity of the accounts are pertinent less as issues of historical truth but more as highlighting of tendencies within the hagiography. They suggest to us what may be the author's spin on the material. Three cases are illustrative. The first concerns Maha Bua's biography of Ajahn Mun. The second is David Chadwick's (2000) appreciative paean to Suzuki Roshi. The third is Sheng Yen's autobiography, *Footprints in the Snow* (2008). Clearly, all of these works are modern productions in an age where little passes without controversy. Nevertheless, the content of the objections is revealing.

Maha Bua has been criticized primarily on his doctrine of the pure mind (see Chap. 7) which some interpret as a covert atman. However, objections have also been raised with respect to his biographical efforts. The biography of the saint put the forest movement on the map, and undoubtedly Maha Bua's efforts have been the primary contributors. While there is no full-length critique, several objections have been voiced. Maha Bua has been characterized as a "Johnny come lately" having practiced with Mun for a relatively brief time and often not living in any proximity with the Master. How did he collect the knowledge he claims of Mun's life? A senior monk of the tradition has disputed Maha Bua's portrait of Mun. To his eyes and ears, Mun did neither sounded nor acted as Maha Bua presents him (dhamma wheel 2009). Whatever the truth, it reminds us that the text is social fact in itself. Maha Bua's Ajahn Mun is the face of the Thai Forest Movement.

The two Ch'an/Zen cases suggest the skewing of hagiography to meet in one instance traditional themes and in the other to align with present day political correctness. Sifu Sheng Yen presents himself as living through privations in a cold New York City winter. A former student finds doubtful his narrative of cold and hunger. Unlike the mountain monks of old and the monk-pilgrims trudging over the roof of the world, Sheng Yen was invited to the West by wealthy Chinese and supported by established

Chinese Buddhist institutions. The Road of Trials demands such episodes of privations. Any monk conscious of his public persona and place in Ch'an memory would insist on such chapters as integral to his story (Lachs 2011).

Likewise, Suzuki Roshi's biography has been challenged but with regard to his reported pacifism under Japanese militarism. An effort to trace the origins of such claims led not to solid sources but to a flat denial by Suzuki's son. As one informant stated it, "We were all peaceniks; against the Vietnamese War." A teacher who like so many accommodated with militarism was not the Zen wanted in the 1960s San Francisco Bay Area (Lachs 2011). What both Sheng Yen and Suzuki's narratives share is a promotion of their traditions and especially its teachers as standing outside the mainstream order and willing to undergo trials for the Great Matter. The criticisms, valid or not, underscore the tradition's stylized self-presentation.

Summary

The lives of the awakened constitute a vast literature whose breath goes beyond our present inquiry. However, in restricting our examination to the primary line of interest of this work, several key points come forth. There are differences between the Zen genre and the Thai compositions. The Zen sect self-consciously develops a "Zen master archetype," while the forest movement relies on the Buddha story and the great arahants as its primary template. As noted, the Ch'an/Zen school narratives include the anti-saint, the rascal type, who flaunts conventional precepts, as well as vital role played by women of wisdom. The character of the anti-saint/rascal and the wise woman in Ch'an/Zen has no easy correlate in the Forest Tradition.

Within the Forest School, the style of the Maha Bua sub-sect contrasts with the Ajahn Chah lineage. Maha Bua's accounts focus on the enlightenment attainment and tend toward the laudatory. Ajahn Chah biographies play down awakening as a specific event and avoid the monk stereotype for a more nuanced and humanistic narrative. The key difference is that the Kammatthana remains closer to the Buddha's story while Ch'an develops a composite archetype of iconoclastic power drawn from Bodhidharma, Huineng, Matsu, and other T'ang era figures. Nevertheless, there are sustaining similarities. First is the cardinal role played by auto/biography in both schools. The critical role of the charismatic figure and a highly personalized approach to the religious project is once again demonstrated. Second, the hagiography becomes a synthesis of the cosmological and the individual by draping the personal story over an archetypical structure. Third, the stories convey not only inspiration and core values but teachings and guidelines for practice via flesh and blood persons. We found that the stories of the awakened reflect the pattern of Campbell's hero/quest and van Gennep's rites of passage. They are a part of a world literature of dramatic quest and triumph whose themes of leave taking, search, trials, and ultimate victory speak to the human condition. Finally, it is worth noting that an endeavor framed as adventure is compelling. What could be more exciting to village youths than to test their manhood in a rite of passage leading to awakening?

References

Amaro, B. (2012). *The life and teachings of Ajahn Chah.* Hertfordshier, UK: Amaravati Publications.

Braverman, A. (2013). *Mud and water.* Somerville, MA: Wisdom.

Breiter, P. (2004). *Venerable father.* New York: Paraview.

Campbell, J. (1972). *Hero with a thousand faces.* New York: Harper Collins.

Chadwick, D. (2000). *Crooked cucumber.* New York: Broadway Books.

Chah, A. (2005). *Everything arises, everything falls away.* Boston, MA: Shambhala.

Chang, G. C. C. (1959). *The practice of Zen.* New York: Harper and Row Publishers.

Dhamma Wheel. (2009). Retrieved April 4, 2015, from http://www.dhammawheel.com/viewtopic.php?t=1000

Dumoulin, H. (2005a). *Zen Buddhism: A history* (Vol. 1). Bloomington, IN: World Wisdom.

Durkheim, E. (2008). *Elementary forms of religious life.* Oxford: Oxford University Press.

Ferguson, A. (2000). *Zen's Chinese heritage.* Boston, MA: Wisdom.

Ferguson, A. (2012). *Tracking Bodhidharma.* Berkeley, CA: Counterpoint Press. Kindle edition.

Foong Thim Leng. (2011). *Lungata Maha Bua's early life.* Retrieved September 5, 2015, from http://thaiforestradition.blogspot.com/p/a-biography-of-lungata-maha-buas-early.html

van Gennep, A. (1960). *Rites of passage.* Chicago: Universtiy of Chicago Press.

Heine, S., & Wright, D. (2010). *Zen masters.* Oxford: Oxford University Press.

Hsu Yun. (1988). *Empty cloud: The autobiography of the Chinese zen master Hsu Yun* (C. Luk, Trans.). Rochester: Empty Cloud Press/Element Books.

Lachs, S. (2011). *When the saints go marching in*. Retrieved May 4, 2015, from http://www.thezensite.com/MainPages/critical_zen.html
Lee, D. (1991). *The autobiography of Phra Ajaan Lee*. Valley Center, CA: Metta Forest Monastery.
Luk, C. (1970). *Ch'an and zen teachings: First series*. Berkeley, CA: Shambhala.
MahaBua(1976). *The VenerablePhraAcariyaMun*,SamudraSakornWatPrajayarangsi.
Maha Bua, B. (2005a). *Arahattamagga, Arahattaphala*. Udon Thani: Forest Dhamma Books.
Maha Bua, B. (2005b). *Wisdom develops samadhi*. Udon Thani: Forest Dhamma Books.
Maha Bua, B. (2006). *Venerable Ajaan Khao Analayo*. Udon Thani: Forest Dhamma Books.
Porter, B. (2009). *Zen baggage*. Berkeley, CA: Counterpoint Press.
Shaw, J. (2014). *Adventures of the mad monk Ji Gong*. Rutland, VT: Tuttle Publishing.
Sheng Yen. (2008). *Footprints in the snow*. New York: Doubleday.
Silaratano, B. (2009). *Mae Chi Kaew*. Udon Thani: Forest Dhamma Books.
Tambiah, S. (1976). *World conqueror and World renouncer*. Cambridge: Cambridge University Press.
Thate, A. (1993). *The autobiography of a forest monk*. Wat Hin Mark Peng: Amarin Printing.
Tiyavanich, K. (1997). *Forest recollections*. Chiang Mai: Silkworm Books.
Waddell, N. (2000). *The unborn*. New York: North Point Press.
Waddell, N. (2011). *Wild ivy: The spiritual autobiography of Zen Master Hakuin*. Boston, MA: Shambhala. Kindle edition.

CHAPTER 7

Pure Mind: The Fifth Noble Truth

Introduction

Turning to doctrine raises our study to an entirely new level. Until this point, we have examined these movements as systems of activity: teaching tactics, practice and path, the recorded lives of their luminaries, and the enlightenment experience. We have approached doctrine in terms of its function with regard to the action of the movements. The soteriological project, the attainability of awakening, and the need for self-cultivation were noted as orientations that give direction and dynamic to the monks and masters of the forests and mountains. Here, we examine doctrine to note how the soteriological possibility is framed and defined. Our primary reference point will continue to be the pragmatic consequences of doctrinal innovations rather than attempting to adjudicate the often ornate intra-Buddhist debates.

Focus will be primarily on the Thai Forest Movement for several reasons. First, because the Forest Movement's emphasis, if not innovation, places it tension with its Theravada brethren. Second, because the major polemical thrust of the movement has been to explicate and defend this doctrinal stance. Third, the teaching of the pure mind moves the forest monks closer to the mountain monks doctrinally but also in their axis of practice. Finally, it can be noted that Zen/Ch'an while displaying creativity in practice and at times in the spin given to Mahayana doctrine

A.R. Lopez, *Buddhist Revivalist Movements*,
DOI 10.1057/978-1-137-54086-7_7

nevertheless is largely consonant with its parent tradition. Mainstream Mahayana gave Zen the ideas it required. Zen gave them immediacy.

Theravada is the Buddhist school of orthodoxy. They are the Thera, the Elders, of the sangha who claim the preservation of the Buddha's original teachings. Nothing is more cardinal to Theravada teachings than the Four Noble Truths and the Three Marks of Existence, dukkha, annica, and anatta. The first Truth and Mark agree on the foundational insight that conditioned existence entails disease. The Truths of course go on to identify the cause, cessation, and way out of suffering. The Marks note two additional qualities of the conditioned world, impermanence and selflessness.

Into the matrix of these interlocking notions, the Forest Tradition introduces what could be termed a Fifth Noble Truth. Maha Bua analyzes the Four Noble Truths into two pairs. The first and second tell us what is the problem, while the third and fourth tell us of a solution to our circumstance. So far so good, Maha Bua seems to say, but what lies beyond the problem and the way to its resolution? The answer is the pure *citta* or mind which could be called the fifth noble truth. As we shall show the Dhamma of the Forest Tradition takes on an immediacy and optimism that critics have often seen lacking in classical Theravada thought. Out of the dark forests of Thailand, the Kammatthana monks tell us that not only is the Unconditioned within reach but that it is none other than the pure, shining heart/mind.

Doctrinal Setting

The concept of pure mind redirects Theravada doctrine not by rejecting mainstream teachings but by an addition that redefines enlightenment and Nibbana not simply as an absence but as positive vibrancy, a radiant presence. Yet the pure mind is more than an add-on to the Noble Truths. The disciple is being directed to search for Nibbana not in the heavens but in this very mind. Any misconception that awakening is annihilation is refuted. Sentience, and a brilliant sentience at that, is held out as the supreme attainment. In the place of a negative lexicon of suffering, impermanence, and no-self are the qualities of the pure mind: permanence and self-nature.

What Ch'an and the Forest Tradition share is a revivalism that advertised itself as a continuation/renewal of Buddhism through uncompromising practice. Bodhidharma's direct pointing was not about a new doctrine or goal, but about the direct pointing itself. Experiential self-cultivation unwaveringly oriented toward liberation is Zen as opposed to the immense translation projects funded by the Emperor. The esteem for translators of the foreign doctrine into Chinese should not be underestimated. Murals and wall hanging depicting the translation team typically show them sitting below an over-sized monarch, except for the chief scribe, usually an Indian monk, placed above the Son of Heaven's head. The location of the main translator illustrates the unsurpassed value of the transcription of Sanskrit into Chinese and the esteem for those who directed the project. In this world of Chinese Buddhism, Bodhidharma unleashes his direct pointing.

Despite the commitment to action, Zen has its doctrinal dimension and is by no means a mindless Buddhism. The manuscripts of Tungshan record Ch'an masters well versed in the sutras and not shy about invoking them to make their point. Shitou, a T'ang dynasty master, leaves us the *Discourse on Unity of Sameness and Difference* which draws on indigenous Chinese thought. Several generations later, Master Dongshan formulates the Theory of the Five Ranks a Soto model for progressive, multidimensional realization. To convey his philosophical insights, Dogen, the thirteenth century revivalist of Japanese Zen, invented new language. On sum, the intention of Zen, however, was not to produce new systematic philosophy as much as to translate pre-existing abstract theory into practice sharpening categories. Like the Forest Masters, Ch'an/Zen employed a language that guided and inspired practice. Terms such as Original Face or the Unborn Buddha Mind imparted an immediacy absent in sutra-based discourses on *tathagatagarbha*, Buddha-nature, or alayavijnana, the storehouse consciousness.

What differed between Ch'an and the Forest Tradition was the former's more heterogeneous doctrinal inheritance. In China, different sects coalesced around different texts. Zen had the luxury of migrating across these texts from the Lankavatara to the Prajnaparamita to the Avatamsaka to the Lotus Sutra without encountering orthodoxy such as exists in the Theravada world. In addition, the philosophical systems imported from India largely supplied Zen with all the intellectual resources needed. Ch'an/Zen was not pointing at something new to Mahayana, it was the directness of its pointing that was new. Zen could afford to be more pragmatic-practice oriented.

The Forest Tradition faced a different task. To give conception to its view demanded assertion of ideas at odds with Theravada orthodoxy. The forest masters were on a different symbolic terrain than their mountain siblings. What these innovations are, how they were defended, how they shifted the axis of practice, and finally how they bring the Forest Tradition into greater alignment with Ch'an/Zen is the story we will turn to in this and the following chapter.

Even as Ch'an separated itself from the sutra-based philosophical schools while invoking them as needed, it did innovate concepts that sharpened its soteriological project. A review of Zen language reveals a lexicon concerned with the description of the fundamental nature of the heart/mind. Terms such as the True Man of No Rank, Your Face Before Your Parents were Born, the Unborn Buddha Mind, the Mind Essence, Original Mind, and Ordinary Mind are synonyms for ultimate reality, but one that is none other than mind itself. The Unconditioned is located in or as the heart/mind. The domain of mind as the locus of transcendence and reality for Zen is repeated in the Forest Tradition's teachings on pure, shining mind. Based on the Yogacara philosophy, Zen could say all is mind, while Theravada would only go so far as to say that all experience involves mind. Nevertheless, both schools agree that mind is central.

From where did this notion of a radiant mind that supersedes the conventional mind derive? What we can rule out is that they influenced or borrowed from one another. There has been suggestion that the ethnic minorities migrating into Thailand from China may have brought such notions. This conjecture is purely speculative with no evidence. Furthermore, we find no presence of such ideas among the hill tribes or their monks. It is unlikely that these animist/Buddhist groups conveyed any sophisticated Pali or Sanskrit thought to Ajahn Mun and his forest brothers. More plausible sources are contemporary trends in Theravada Thai Buddhism. While the rationalizer of modern Thai Buddhism, Buddhadasa, did translate Huang Po and Huineng into Thai. Buddhadasa is not by lineage part of the Forest Order, and the idea of pure mind finds expression among forest teachers prior to the publication of Buddhadasa's translations. Furthermore, Buddhadasa's writings do not contain explicit mention of Maha Bua's pure mind. Also, we have no evidence that Forest teachers were conversant with Thai translated Zen. Ch'an's cultural and geographic distance from Pali Buddhism precludes any direct north to south influence.

The lone possible exception to the absence of direct contact is Ajahn Chah. He did read Mahayana texts and Zen works. However, this was only because students had noticed a marked similarity in the style and content of Chah's teaching. For example, a former monk of Chah's had taken up

Zen practice in the USA. On a return visit, he asked Chah's opinion on Zen teachings and provided him with samples to peruse. To the student's delight and amazement, Chah grasped their meaning, elaborated their implications, and noted their similarity with his own approach (Brieter 2012). Yet here we are dealing with a matured teacher who is only reflecting on Zen and his teachings. Zen was not part of Chah's formative world. So what is the source of Chah's teaching on pure, radiant Original Mind, or as he would say in Thai, *citta tamadah,* natural mind? Chah tells us. The idea that there is a pure mind beyond the defilements was a critical learning he gained from his brief contact with Ajahn Mun (Jayasaro 2008, p. 13).

The Thai Forest doctrine of Original Mind seems to have derived from and finds support from two sources, respectively: anonymous aranyawaassii and the Pali Canon itself. The latter confirms the doctrine, but nowhere does Mun suggest he acquired this learning from a text. Practice itself is guided by teachers who are senior on the path and Mun did search out forest ascetics. Mun does not identify any monks by name, yet his dogged pursuit of forest teachers across Thailand and even into Burma suggests that he may have heard of a pure mind teaching at first from others. The direct transmission seems to have been from master to master-to-be and verified by Mun's practice/realization. How widespread this teaching may have been in forest circles we do not know. Scripture afforded adjunctive confirmation, which Forest Masters liberally quoted and some modern scholars have seconded.

Pure Mind in the Forest Tradition

The teaching of a pure radiant mind appears in the earliest discourses of the Thai Forest Movement. Although Mun's written work is sparse, he conveys the existence of this mind as central to a practice dedicated to liberation. In his poem *The Ballad of the Liberation from the Five Khandas,* Mun tells us in a Zen like phrasing that "Knowing not knowing: that's the method for the heart." With the halting of mental productions and their consequent problems, all that remains is the "Primal Mind, true and unchanging" (Mun 1995a). This is the teaching that is unique to the Forest Tradition and revolutionized the young Ajahn Chah understanding of dhamma.

The second generation also speaks of this mind, with different teachers using slightly different terms, but all describing a sentience that does not partake of two of the Three Marks of Existence, dis-ease, and impermanence, but is characterized by ultimate release, permanence, and a true self-nature. In his work *The Craft of the Heart*, Ajahn Lee leads the reader through the entire path culminating in the realization of the pure mind. Lee is unequivocal about the place of the pure mind in Buddhism. "Nibbana is nothing else but this ordinary heart," he asserts (1994, p. 160). Ajahn Chah repeatedly stresses the inherent purity of the mind. Despite all our defilements, "It is intrinsically pure," he assures us. Despite our agitation, it is "already unmoving and peaceful." He agrees with Lee in that this mind is the goal of the Path. "Our practice is to see the Original Mind." He concludes that "Just this is the aim of all this difficult practice we put ourselves through." Chah's approach emphasizes both the ultimate value and the already existent nature of the pure mind (Chah 2002, p. 116).

Ajahn Thate, who like Chah had Western students and traveled to the West, speaks of an "innate purity and colourlessness" of the heart/mind. In a teaching entitled *Only the World Ends (2007),* Thate contrasts the conditioned world with the unconditioned or the radiant mind. He uses the metaphor of water to explain the apparent impurity of mind and its enduring condition of purity. Even though water can be colored by introducing external substances, its fundamental nature remains. The defilements therefore are conditional, even accidental, to the inherent or original nature of mind (2007, p. 82).

Of all the heirs of Mun, the one who would expound most explicitly on the doctrine of radiant mind is Maha Bua. He carefully clarifies its nature, distinguishing it from any hidden atman or self-entity despite its permanent nature. The discovery of this mind is placed in the context of his personal practice and marks, without explicitly saying so, his entrance into arahantship. Yet most significant is his discovery that radiance is ignorance or *avijja*. In fact it is the ultimate seduction on the path. The radiance needs to be penetrated to reveal the pure, true citta, which is beyond owning and possessing (1987, p. 230). Maha Bua's formulations will be further explored in relation to Mahayana doctrines of *tathagatagarbha* and *alayavijanna*, and the relationship between the task of practice and an already existing pure nature.

The Nature of Pure Mind

The first and second generation of Forest Masters expounded the doctrine of a luminous mind, but what is its nature? A review of how this mind is characterized by the Forest Teachers produces a set of aspects which when taken together sketch a coherent phenomenon out of what otherwise might be considered an ineffable mystery or mere idiosyncratic idea. The following qualities appear repeatedly in Thai forest teachings:

Sentience: The mind has the capacity of knowing. In fact, it would be more accurate to say that this mind is the capacity for knowing. Maha Bua observes that everything disappears but that which knows the disappearing does not (1987, p. 237). The capacity for knowing is perhaps the cardinal quality for it categorically refutes any notion of annihilation or Nibbana as a dead zone. Of course, this consciousness is to be distinguished from any sense-based or thought-involved cognition and could be understood of as the very essence of awareness.

Radiance/luminosity: Radiance and luminosity further convey the quality of sentience, as well as purity. If sentience is central to the description, then radiance and luminosity suggest awakeness with an attractive luster. This is not merely a profane cognitive faculty but one that shines with sacred. Finally, the terminology connects to the early texts where the Buddha himself uses the term luminous or radiant mind, *prabhassara citta*. When the mind is unobscured by defilements, it is radiant. On this point, both Pali passages and the forest teachings seem agree.

Inherent/innate: This mind is in no way produced (or destroyed), constructed, derivative, or dependent on causes and conditions. The pure mind is primary or foundational not in temporal sequence but existentially. It is priori to time and space and requires no human or non-human agency to bring it into being. Apart from this august status, the pure mind as inherent and innate is always present to the path and the practitioner. There is an active available aspect. Ajahn Chah suggests as much in his witty and intriguing simile when he says "You're riding on a horse and asking where's the horse?" (2004, p. 154). We are already riding on the pure mind. We are carried by it, in a manner similar to Zen's contention that we a supported by the way seeking mind or that our Original Mind is right before our eyes or under our feet.

Purity: The dichotomy of purity/defilement so central to Theravada thought finds its resolution in the pure mind which is not only free of the defilements of the mind but in its innate purity which supersedes polarities of conventional right and wrong. The purity of the mind seems to be a function of its non-dependence on the conditioned world. Here, purity is not just moral cleanliness in a conventional sense, but a transcendent, ontological purity of non-entanglement in the conditioned world. The pure mind is not so much described in terms of moral agency as much as an unmitigated phenomenon, such as pure water.

Permanence/constancy: The luminous mind does not waver, fluctuate, and certainly does not arise or fall away. Even as the world, the universe, ends the luminous mind is. Only the world ends Ajahn Thate reminds us. Even the Theravada concern with impermanence ceases with the timeless, pure mind. Maha Bua, for example, discerns the subtle wavering of Mae Chee Kaew's mental state and rejects her attainment. The pure mind is utterly unchanging, like Nibbana.

Natural/ordinary: For all its extraordinary qualities that raise pure mind above the world of causes and conditions, it is deemed ordinary and natural. Repeatedly, forest teachers speak of seeing the world as it is or as it has always been as the cognitive mode of this mind. Because it is innate/inherent, it is uncontrived that is natural both in its condition and its functioning. In Thai, it is *citta tamadah*: natural mind. The natural and the ordinary are implied by the innate and the inherent aspects.

Ultimate release: The notion of a pure mind is not only of immense philosophical and perhaps psychological implication, it is primarily a concept of soteriological significance. The attainment of this mind is the great release from stress. With the practitioner's entrance into the luminous mind, there is the laying down of the burden. There is the temporary stopping of the khandas, the psychophysical complex of the personality. The whirling of the conventional mind is replaced by a deep peace. Dukkha is vanquished. Dhamma is directly known. The uncovering of the luminous mind is what all our hard practice is for, Ajahn Chah reminds us.

No-self: Finally, the pure mind is free of any trace of self. Maha Bua is particularly emphatic on this point. If there is any trace of an "owner" of an experience, it is a condition of *avijja*, ignorance. In short, there can be no "my" enlightenment. Awakening is without an owner. Maha Bua documents his ruthless eradication of any trace of self in his quest. Although possessed of a self-nature and permanence, it is not a self. While exhibiting the qualities just enumerated, the pure mind is utterly non-personal and devoid of any theistic suggestion. It is empty, *sunna* (Pali). Empty of what? Pure mind is empty of any trace of a self.

A review of these characteristics shows a match between pure mind and Nibbana. The former has a mentalist feel that is not evident in Nibbana which can be and has been construed as a place, realm, or ontological state. A careful review of both concepts shows mutual identity. Nibbana and pure mind share the cardinal features that have traditionally marked the Buddhist absolute. Both are permanent, free of suffering, free of defilements, and not-self. The *Milindapanha's* soaring description of Nibbana applies equally to the pure mind. We are told that Nibbana is like space, like a mountain characterized by transcendent purity, unshakable permanent, and unsurpassed in loftiness (Mendis 1993, pp. 132–133). Each of these qualities and their metaphors applies to the pure mind celebrated by the Thai Forest masters.

By Any Name

Like the highest in many religious traditions, this mind is known by many names. Given its multiple characteristics, it is to be expected that different teachers, on different occasions, will delineate different features. Terms such as Original Mind denote its inherent/innate status. The uncontrived aspect is communicated as Natural Mind. Freedom from defilement is the Pure Mind. Radiant or luminous mind makes visible the invisible knowing aspect. *Poo ruu* (Th.) or the one who knows gives personal and colloquial shading to the sentient capacity. Perhaps the most intimate of terms is simply Heart (*hua jai*, Th.).

The last two terms, Poo Ruu and Heart, warrant further comment. The One Who Knows can be mistaken to imply an inner person, perhaps a homunculus, who cognizes. As Amaro and Pasanno (2009, p. 192) indicate, this would be an erroneous personification of a Thai expression that is used in a wide range of situations and should not be taken as a philosophical position or psychological entity. It might be more accurate but less expressive to say *that* which knows. "Heart" might conjure, especially to the Western ear, an organ of affect. To a Thai, the meaning is different. As with the ancient Greeks, the mind for the Thai locates in the center of the chest. The word is hua jai, literally head heart. To understand in Thai, for instance, is to *kao jai*, to enter the heart. This heart may be distinguished from the *jit jai* or thinking heart. Ajahn Thate (1993, p. 84) makes the point that heart here also means central as in the heart of the matter. Both terms, Poo Ruu and Heart, give a more human tone to what might otherwise be heard as a distant, airy concept.

What generates and guides the production and use of a variety of terms is, of course, the imperative of effective communication. The Forest School is a teaching/practice tradition whose interest is in directing and inspiring the disciple. Neat logically tight categories, such as those in the Abhidhamma texts, that would be the requirement of scholar monks, are unnecessary for the practitioner in the forest.

Teaching and Scriptural Defenses

The founding generation introduces the notion of a mind other than the conventional citta. The second generation, Mun's direct students (Maha Bua, Ajahn Chah, Analayo, Dhammadaro Lee, Thate), promulgates the doctrine so that it comes to be the signature of the Thai Forest Tradition. What we can loosely designate as the third generation grounds the teaching in scripture and through their commentary sharpens the doctrine, unpacking its larger meaning, and defending it against critics.

The teachings of the second generation are predominantly oral which usually treat classical Dhamma themes but with the teaching of radiant mind included. For instance, Ajahn Chah describes pure awareness in a talk to a dying lay disciple. He dramatically and intimately terms this knowing of pure awareness as "meeting the Buddha" (2002, p. 326). In this discourse, the Buddha is no longer a man who walked northern India but the principle of what is a Buddha, pure mind that is in reach of each of us. The similarity with Mahayana Zen doctrine is evident. In the Mahayana, the Buddha is the *sho* or principle, the fundamental reality, with a human name. T'ang Dynasty masters spoke of Buddha Mind. When the Japanese master, Dogen, would switch a more religious mode of discourse, the word 'Buddha' would appear. In his teaching at the death bed of a disciple, Chah instructs her to meet the real Buddha. In effect he saying, "Shining mind is Buddha" (2002, p. 326). Dhamadharo Lee leaves behind several written documents that give prominence to the pure mind teaching. In both his *Craft of the Heart* and in his autobiography, the teachings of his master, Ajahn Mun, on the radiant heart are presented. "Nibbana," he tells us, "is nothing but this ordinary mind" (1994, p. 160). The phrase repeats, unknowingly, the essential teaching of Zen Master Mazu from more than a thousand years prior: Ordinary Mind is Buddha. Like his brother monk, Ajahn Chah, the equation of pure mind with the ultimate (Buddha, Nibbana) is unequivocal.

Second-generation teachings are truths from the forest. They are not commentarial, scholarly defenses of what to a more mainstream Theravada is heterodoxy. The task of dhamma combat on the intellectual stage falls to and has been picked up by the third generation. Several monks have taken up the task of defending forest doctrine. In particular, we can look to two of Ajahn Chah's students, Amaro and Pasanno Bhikkhus and a disciple of Fuang Jotiko, Thanissaro Bhikkhu. Notably, all three monks are Westerners with significant formal education. More than Thai monks who seem content to accept and practice the teachings, Westerners exhibit a great need to explain, defend, and propagate the word.

Two works in particular address the nature of mind and Nibbana. First is an anthology entitled *The Island* (2009), a classical metaphor for Nibbana, collated by Amaro and Pasanno. Second is a slim but focused volume by Thanissaro, interestingly entitled *The Mind Like Fire Unbound* (1993). Sections from the later are included in the former work. The expressed purpose of Amaro and Pasanno's work is to bring Nibbana back into the center of Buddhist consideration. In a private conversation, Amaro asserted, "That's what this is all about!" Liberation, the attainment of Nibbana, is the *raison d'etre* of Buddhism. The soteriological project is primary. Theravada has been criticized that they only have Noble Truths addressing the conditioned world: suffering, its cause, its end, and a path, but no transcendent truth. In conversation, Amaro indicated that it was precisely to correct this misrepresentation that *The Island* was produced.

The Island includes selections on Nibbana from the Pali Canon that confirm it as a central theme and no mere mystical after thought. Lists of synonyms demonstrate Nibbana as a recurrent and fully developed notion in the Canon and to offer the reader a more complete treatment of a teaching that has been too often ignored. The authors also range outside of the Canon to find corollaries in the Mahayana and indeed beyond Buddhism. They seek to establish that Nibbana is one of the core truths of Buddhism and that it is a positive, attractive reality. Vital to their argument is that Nibbana is not extinction. This contention goes further than complimentary imagery and attractive metaphors. The assertion of a positive Nibbana turns on the existence of an enduring, unconditioned mode of discrimination or consciousness. In short, that there is a pure, radiant mind that remains with the falling away of all attachments and defilements. The teaching of the forest masters conveys a Nibbana that is not extinction but the existence of an intimate immediacy: the natural, ordinary mind itself.

To demonstrate this, a key metaphor of the Buddha that has been construed to mean extinction requires explanation and clarification. The image/metaphor is "fire" and it is going to ground upon liberation. Fire is employed by the Buddha in several ways in the Pali Canon, *The Fire Sutta*, generally identified as the Buddha's third sermon, is the fire of attachment and the passions. However, the Buddha also uses fire as the metaphor for consciousness. This allows him to suggest a condition of agitation and suffering, but also in a liberated state. Thanissaro's *Mind Like Fire Unbound* explores this element and its dynamics in classical Indian thought in order to unpack the Buddha's true meaning. The discussion, which is repeated in Amaro and Pasanno's work, seeks to rescue Theravada Buddhism from any annihilationist and nihilist interpretations and at the same time align early Buddhism with the teachings of the Forest Masters.

The thesis is simple: conscious/mind does not cease to exist with the attainment of Nibbana. In fact, this mode of unsupported discrimination or free, unqualified consciousness *is* Nibbana, the timeless, unconditioned, and uncreated. The argument rests on an understanding of the early Vedic and Brahmanical model of the elements. Historical investigations uncover an ancient model in which elements are activated under certain conditions, but when inactive do not cease to exist but return to a "non-manifestive" state, their latent home ground. Understood against the physics of his era, the Buddha's words take on new meaning. Speaking to an audience which presumably shared the early Indian conception of elements and their functioning, the Buddha's words convey a message different from his later interpreters. The Awakened One speaks of a "Consciousness which is non-manifestive, endless, and lustrous on all sides." Here, consciousness exists beyond its usual manifest condition as when agitated by a dependence on conditions. The notion of a consciousness independent of conditions is made explicit in the Buddha's famed reference to a "consciousness [that] is signless, boundless, all-luminous, where earth, water, fire, and air find no footing" (Thanissaro 1993, p. 30). Independent of the conditioned world of elements, that is "finding no footing," there is a luminous consciousness that is infinite in dimension. Thanissaro translates another passage as, "Like a flame going out was the liberation of awareness" (p. 12). Understood within the classical Indian physics, this indicates a liberation of awareness not its annihilation, and its return to a non-manifestive state that is free of worldly limitations. The references to luster link the Pali passages to the shining citta spoken of by the Forest Masters.

The monks Pasanno and Amaro cite Thanissaro's argument in their extended investigation of Nibbana. Moving beyond the metaphor of fire, they document the shining mind and its associated form of consciousness as Nibbana itself. Aside from the textual intricacies of their presentation, what is critical is their apologetics for the Forest Movement. *The Island* is not merely a scholarly production; it is a project to correct Theravada thought and to defend the Thai Forest teaching of the luminous mind. This is conveyed in their introduction and in conversation with the researcher. As Ajahn Pasanno states, contrary to popular belief that the Buddha had little to say regards the absolute, "the Buddha *did* say a great deal about Nibbana" (italics original) (p. IX).

The Buddha's non-manifestive consciousness is also the unsupported consciousness or discernment that appears in the Pali Suttas. The idea of a form of consciousness that is not a function of sense data or thought process, and is unsupported by the conditional, affirms the notion of pure mind. What else could pure mind be other than a reservoir of pure consciousness? We need to be clear that the pure mind is not some thinking entity like a cosmic headquarters dispensing commands to the universe. Rather it is, to use the title of Ajahn Jotiko's book, *Awareness Itself* (1999). The third generation of forest teachers links this consciousness beyond conventional cognition to create a triangle of equilateral meanings: Pure Mind = Nibbana = Unsupported Consciousness.

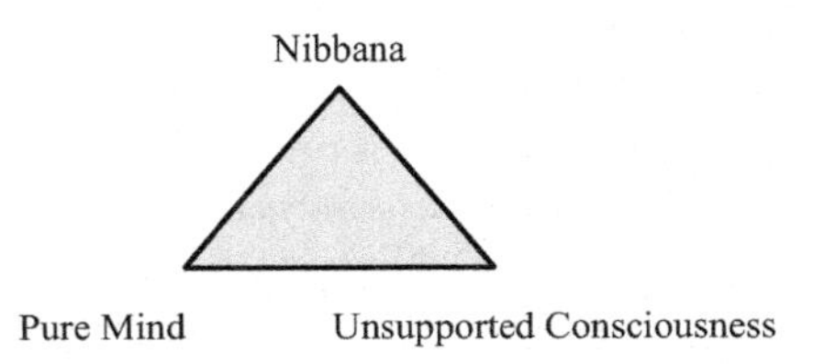

Early Suttas and the Abhidhamma

The forest monks are not alone in their promotion of a so-called Fifth Noble Truth. They receive ample support from scholars, both lay and ordained. Recent scholarship has made a distinction between the Pali Canon and the Theravada tradition. Historically, this is supported by the recognition that there were numerous sects in North India who claimed the Canon as their doctrinal foundation. In other words, the Theravada interpretation is just one reading of the texts. Other doctrines may lurk in the palm leaves, waiting for a discerning eye and ear.

Peter Harvey is one such explorer. In his *The Selfless Mind* (1995) and related articles, Harvey discovers a Dhamma of nibbanic consciousness and the pure radiant mind. The passages that he marshals in support of his selfless mind are Forest Teacher favorites. Harvey briefly speculates that the Thai Forest Masters seem to be propounding what his textual investigations unearth. Our investigation confirms his hunch. The suttas support the Thai Forest Teachings not only in their discussion of Nibbana, as Pasanno and Amaro demonstrate, but also with two critical and related teachings. The first is the already mentioned unsupported consciousness/ discernment. Second is the bhavanaga consciousness that appears in the

Abhidhamma literature. Turning to the term vinnana, translated as consciousness or discernment, the Buddha expounds on this mode of awareness as follows:

> That unsupported (apatitthitam) discernment has no increase is without constructing activities, released. (Harvey 1995, p. 126)

The Buddha goes on to credit this release with steadfastness, contentment, and the attainment of Nibbana. The very qualities of unsupported consciousness are bundled with Nibbana. Like the Ultimate, this mode of knowing is beyond measure (no increase/decrease), without construction, and unwaveringly peaceful. One is reminded of the *Udana* passage where the Buddha speaks of Nibbana as the Unborn and the Uncreated. Here, he attributes similar qualities to the apatithitam consciousness. Again we hear that this consciousness is non-manifestive, infinite, and shining in every respect (*vinnanam anidasanam anantam sabbato-pabham*). Elsewhere, the Buddha identifies Nibbana as unsupported and without maintenance, further linking the mode of consciousness with transcendent reality (1995, p. 200).

While the apatithitam consciousness undoubtedly has an affinity with Nibbana and is the vehicle for the perception and attainment of Nibbana, some scholars (Harvey 1995; Johansson 1969) assert an even more unqualified relationship between this mode of consciousness and Nibbana. For Harvey, Nibbana is a form of consciousness (even as it can be an object of discernment). Harvey terms this nibbanic-consciousness, while Johansson pictures layers of mind with the deepest being Nibbana itself. In Harvey's reading, this transcendent consciousness is like Nibbana, infinite, because it is objectless consciousness. A consciousness without an object, it is the subject of the Buddha's metaphor of beam of light that falls on no object and therefore continues without stopping. It has no end, and therefore, counter intuitively, is "accessible from all round." It is everywhere, unlimited (1995, p. 202).

This transcendent reality is always "there" even if unrecognized by the deluded. Here is a major theme developed in Mahayana and Zen thought. Of equal importance is the interpretation that a consciousness/discernment, unsupported/objectless/infinite/luminous continues at death and during the suspension of the personality and its cognitive activity in life is Nibbana. Nibbana interpreted as a mode of consciousness is at odds with the conventional Theravada position yet, according to some scholars, finds support in the early suttas. The Forest Tradition can therefore credibly claim to be the heirs of the Buddha not only in their rigorous practice but also in their understanding of the nature of the ultimate Dhamma.

A second concept along with the unsupported consciousness invites comparison with the Forest Teachings. While having antecedents in the Sutta Pitaka, the concept of *bhavanga* is found in the Abhidhamma. In the course of developing a more precise psychology of perception that would provide a model of the micro-process of the arising and falling away of mind moments in the perceptual act, Abhidhamma authors identified a ground or resting state of consciousness from which the mind moves and to which it returns at the beginning and end of a perceptual cycle. Nascent forms of the bhavanga idea can be found in the suttas in the form of mula or root consciousness. The idea of bhavanga indirectly supports the notion of unsupported discernment in that it is clearly also a mode of consciousness that is neither sensory nor thought based. The argument of Thanissaro and others that the Indian concept of fire as returning to its home-ground when not manifest suggests the idea of a ground consciousness, bhavanga.

However, bhavanga is related to the Forest Teaching in two additional ways: the phenomenological, descriptively, and by its role in awakening. The adjectives that apply to the bhavanga consciousness replicate the language use by forest masters to describe radiant mind. The bhavanga is radiant the Milindapanha tells us. A simile of the sun is used to convey its self-radiant nature. Elsewhere, it is termed *pakati* (natural) citta (mind), a phrasing we have encountered with the forest monks. Buddhaghosa links the radiant citta with the naturally pure bhavanga citta (Harvey 1995, p. 170). Although the bhavanga is luminous, like the radiant mind, it is only in a proximate relationship to awakening. Bhavanga is not Nibbana, but it is a close neighbor. For all its radiance, it still has latent tendencies to defilements. The bhavanga can and does typically pursue and misconstrue perceptions. Awakening it seems entails another step in which the bhavanga is purified through wisdom-awareness, panna, and comes to shine through all layers of the mind. This formulation seems to parallel Maha Bua's insistence that radiant mind is still avijja, ignorance, and that only the pure mind can be equated with true enlightenment.

Clearly, the above arguments hinge on interpretations of scattered passages out of which a coherent understanding is being fashioned. There are counter arguments. Our intent is not to determine the correct interpretation. What is vital for our study is that the Thai forest teachings are in tension with mainstream Theravada thought, but garner supportive interpretations from the suttas and the Abhidhamma not only from their own monks but from independent scholars as well. A second emergent theme to be explored is that in taking a step away from the scholar monks of their own denomination, the forest masters have moved unintentionally closer to their mountain dwelling Mahayana brothers in the Far East.

Mainstream Theravada Reactions

The Forest Tradition has felt a need to assert its core teaching on pure mind not simply because of the centrality of the doctrine but because it departs from the conventional presentations of dhamma within the Theravada. A casual review of basic texts and introductory overviews to Theravada Buddhism find no reference to the luminous pure mind of the forest. A classic exposition of Theravada Buddhism, *What the Buddha Taught* by Rahula Walpola (1978), presents the Four Noble Truths and unpacks the Eightfold Path but no pure mind or full discussion of Nibbana. Arguably, the preeminent scholar monk of contemporary Thailand, Payutto, has multiple volumes on kamma, co-dependent origination, samadhi, and other expected subjects but no pure mind. The issue is not merely one of omission. The idea of pure mind or nibbanic consciousness is not merely overlooked, it is rejected as smuggling in a self-concept and an eternalism that are anathema to Theravada Buddhism. Conventional Theravada thought restricts its understanding of consciousness to the six conditioned types: the five senses and the mind. Consciousness is seen as impermanent, arising with each perception and then disbanding. Nibbana is, of course, acknowledged but not identified with any transcendent form of consciousness or mind.

The dispute is not merely academic, but goes to the heart of Buddhist liberation. The debate has hit the internet with practitioners posting question about a teaching that is at odds with their understanding of Theravada, "This idea of a base level citta is like a Hindu atman," a monk assure a lay enquirer. He concludes that this teaching "deviates vastly from the Buddha's words in the Nikayas." Although not wanting to directly attack the forest masters, allowing for a misunderstanding of their words, their teachings are in substance unequivocally dismissed (Dhamma wheel 2009). Ajahn Sujato after a meticulous analysis of the word "vijnana" concludes emphatically that, "Nibbana is not consciousness" (http://sujato.wordpress.com).

Whether by omission or blunt repudiation, the Forest tradition clearly faces a need to assert, and defend its position within the Theravada. There is no more damning accusation than being a crypto-Hindu, a status worse than being a Mahayanist. Against such charges, the Thai Forest Tradition's holding to a pure mind doctrine demands defense by third-generation forest teachers such as Amaro, Pasanno, and Thanissaro. The status of pure mind, unsupported consciousness, and Nibbana are critical for the forest teacher's soteriological project. The failure to flush out these teachings in a convincing manner is to deprive the practice of its goal and to strip their teachers of enlightenment. If Nibbana is not a form of mind/discernment, then what is it? To the forest teachers, this leaves Nibbana as an obscure metaphysic, certainly a very un-Buddhist proposition.

Mahayana and Ch'an/Zen on Pure Mind

If the forest teachers can substantially acquit themselves of the crime of injecting a Hindu atman into Buddhadhamma, then the charge of being crypto-Mahayanists is still extant. As we have seen, there is little likelihood that forest dhamma owes anything to direct contact with the Mahayana. Nevertheless, on the point of pure mind/radiant mind, the Forest Tradition and Ch'an/Zen agree. The forest and the mountain without seeking to do so find common terrain. The affinity between Ch'an/Zen and the Thai Forest Movement may be traced back to the early suttas and the Sanskrit/Chinese Agamas. Studies of these early texts suggest that they contain in seed form the ideas of the Mahayana that would later sprout as their sutras and philosophical schools. The Pali suttas, in this interpretation, contain in a nascent state key Mahayana ideas. Ch'an/Zen would then hone these theoretical notions into practice-oriented concepts. While the textual sources might be different, their content suggests a similar creed.

Three core Mahayana ideas find rough corollaries with Pali Sutra teachings. *Tathagatagarbha*, *bodhicitta*, and *alayavijnana* are foundational to both Mahayana in general and to Ch'an/Zen in particular. Tathagatagarbha, famously if unfortunately translated as Buddha nature, is better termed the womb or matrix, garbha, of realization. The notion is not static but undergoes transformation in the Mahayana with perhaps the critical shift for the Zen world being Dogen's reading that we do not have Buddha nature but *are* Buddha nature. The earlier concept affirms our potential for awakening while Dogen declares it to be our ontological condition. We are not just an aggregate of defilements. The problem is Buddhahood is so close we overlook it like the rider on Chah's horse.

Bodhicitta, literally the awakening mind, establishes an innate relationship with liberation. There is within the practitioner's mind an impulse toward enlightenment. In Zen terms this our way-seeking-mind. Arousing bodhicitta provides the motivational fuel for walking the path of the bodhisattva. Both of these teachings impart a positive view of our nature and intimately available support for the journey. Broadly, this performs a similar function to the notion of an inherent pure mind in its uplifting and supportive role, yet the commonality goes further.

Descriptive passages of tathagatagarbha/bodhicitta and the Pali/Forest pure and shining mind share imagery and are conceptually close. We are told of an entity of purity and luminosity. Several of the pre-Mahayana schools seem to propose precursors of the Buddha nature/bodhicitta idea. One of the earliest presentations of the bodhicitta is found in the *Atashakasrika Sutra* circa 100 B.C.E. of *Prajnaparamita* literature which speaks of "a citta that is no citta, since it is by nature brightly shining" (Harvey 1995, p. 175). Mun uses the exact phrase when he speaks of the mind that is no mind. A root text of Ch'an/Zen, the *Lankavatara Sutra*, offers lavish descriptions of the tathagatagarbha which recalls the shining mind of the Pali Suttas and the forest masters. This matrix of enlightenment is "by nature brightly shining and pure." It is hailed as "naturally pure" and "originally pure." Texts such as the *Srimala-devi Simhananda* and the *Ratnagotra-vibhaga Sutras* replicate and extend these descriptions of a "true nature" that is the basis of all existence (Harvey 1995, p. 218).

The *Lankavatara Sutra* also introduces a third Mahayana idea that bears on Pali Abhidhamma teachings: *alayavijnana* or the storehouse consciousness. The alayavijnana is the deepest layer of mind in the Yogacara psychology system, lying below any sense of "I" but being the ground from which the entire personality is constructed. The functional similarity with the Abhidhamma bhavanga is obvious. Both are a kind of continuum of karmic potential but also a basis for awakening. Of particular note is that Ch'an/Zen identified the alayavijnana with buddha nature if it is transformed into the Great Mirror Wisdom of enlightenment. Like the bhavanga, the alaya posits a transpersonal mode of consciousness that suggests the pure shining mind, but is not the pure mind (p. 176).

What our investigation reveals is that both Zen and the Forest Tradition have textual support for the proposition that there is a pure mind whose realization is liberation. Like the Forest Movement, Ch'an/Zen was less concerned with systemic elaboration than with guidance in practice. The sutras may have provided theoretical legitimacy, but the ancestors and masters developed their own labels to point out the goal of practice. Bodhidharma declares a direct pointing at the Mind and the attainment of Buddhahood. The two are equated and his followers are identified as the sect that "meditates on the mind." The practice of the early Ch'an ancestors is *kan hsin*, looking into the mind. Looking into the mind-ground, looking into the mind-source, or looking into the sourceless source are instructions that confirm a basic level of mind as the target of practice.

Zen is replete with innovative names for mind each revealing a different aspect. Mazu of the T'ang Dynasty propounds his famous Ordinary Mind. Linchi personalizes this mind as the Man of No Rank (Watson 1993, p. xxii). Both Chinese and Japanese masters speak of the Original Face before your parents were born. Master Bankie talks of the Unborn Buddha Mind (Waddell 2000). What is consistent is the reference to a mind whose realization is enlightenment. While this mind is, of course, not the conventional citta, we can nonetheless identify a "mentalization" of the goal. Both movements can argue, with textual support, that Nibbana/Buddhahood is a pure mind, innate, and shining. While spoken in different tongues, Thai, Chinese, and Japanese, and from different denominations, their vocabularies replicate one another: Ordinary Mind, Original Mind, Natural Mind.

Images of shining and luminosity also abound in Ch'an/Zen. Han Shan, as we have already noted, speaks of his awakening in terms of "one great Illuminating Whole—omnipresent, lucid, perfect, and serene." He tells of a mind "Bright as moonlight mirrored in the white snow" (Chang 1959, p. 151). Keizan, a descendent of Dogen in fourteenth century Japan, entitles his classic the *Transmission of the Light* (Cook 1996). This light we are told by Ta-chu, Dogen, and others is the "inexhaustible treasury." This is our true inheritance that Ta-sui proclaims to be "originally pure and endowed with myriad virtues" (Cleary 1997b p. 12). The phrasing of the teachers of the mountains and the forests are nearly interchangeable.

The Purity Conundrum

Purity can be a problem. Purity is problematic in two regards. First is the existence of defilements. Second is the rationale for practice. The questions raised are classic, but here we are less interested in the adequacy of their proposed resolutions and more fascinated that near identical problems and solutions appear independently in Mahayana Zen and the Theravada Forest School. With the emergence of a doctrine of original enlightenment, Mahayana Buddhism had to confront two problems. If we are Buddhas, why do we behave in foolish unwholesome ways and suffer? Furthermore, if we are originally enlightened why practice? Mahayana and Zen do not deny the existence of unwholesome tendencies or the requirement of practice. The first problem echoes what theistic traditions call theodicy. How do we account for evil in the creation of an all good god? The more radical the notion of Buddha nature became in the Mahayana the more this problem came to the foreground.

With Dogen's reformulation of the doctrine that we have Buddha nature to that we *are* Buddha nature made a resolution of the apparent contradiction all the more imperative. In fact, it is the living koan that drives Dogen to China and eventually to his teacher Ju-ching. If we are already Buddhas, why all this arduous practice? Dogen concludes famously, "We don't practice to become Buddha. We practice because we *are* Buddha." Sheng Yen retort to a questioning student echoes Dogen with a subtle twist. "If you are a Buddha you are practicing." The union of practice and goal is the Ch'an/Zen solution, especially within the Caodong school.

The Forest Tradition, quite independently of their Zen brothers and sisters, comes to the same logical difficulty of the existence of purity, defilements, and practice. How do we posit a pure mind in world of defilements? Ajahn Chah offers a solution by stressing, in keeping with the suttas, the defilements are adventitious not inherent to the pure mind. He cites the Buddha of the *Vatthupama Sutta*. The Pali words, *agantukehi malehi*, can be translated as adventitious, non-essential, to the mind itself. Chah's argument is in agreement with his teacher Ajahn Mun. Among the fragments of teaching that survive his life is a talk on *thitibhutam*, the primal mind. Mun indicates that this mind is utterly pure and in a spiritual sense neutral in that it can unfold either with wisdom or defilements. The relationship is like that of clouds and the Sun. The Sun remains even when hidden by clouds. The metaphor is identical to a favorite image used in Mahayana Schools (Mun 1995b).

Still it leaves questions as to why do they cover the mind. And of course why do we stick on them? Maha Bua takes on the issue more directly. He introduces a critical conceptual refinement. Maha Bua states the issue as why if the original mind is pure how and why does a world of poisons arise and persist? If the mind is pure, why does it fall for unwholesomeness? Maha Bua makes a critical distinction between radiant mind and pure mind. The radiant mind while apparently innate and fundamental is not pure. In fact, as Maha Bua forcefully points out, the radiant mind is the supreme hideout of avijja, ignorance. According to Maha Bua, ignorance, contrary to our presumptions, is dazzling. He describes ignorance as "the most alluring Miss Universe the world has ever seen." True attainment demands a penetration of the point of radiance. As previously identified in our examination of the enlightenment experience, any trace of an owner, a doer, or localization must be shattered. Only in breaking through the radiant mind and the falling away of avijja is the pure mind revealed. This attainment demands the focused and cutting function of wisdom. Only then is the pure mind revealed. At this point, he invokes scriptural authority. The Buddha refers to *abhassaramidam cittam bhikkhave*, which means radiant mind, and this Maha Bua declares to be "absolutely correct" (1976, p. 174). For, as he argues, if the radiant mind was already pure, there would be no need to purify it. We are born with radiance but not purity. The later requires practice.

To Maha Bua, this distinction between radiant mind and pure mind resolves the conundrum of purity. Regardless of whatever residual arguments may remain, what is pertinent for this study is that Maha Bua's resolution replicates the position of the Abhidhamma with regard to the impurity of bhavanga consciousness and also parallels Mahayana understanding. The bhavanga is pure only in that it is ethically neutral, it can move in either direction. Yet the bhavanga is not immunized from defilements. The alayavijnana is similarly in need of work as it carries the seeds of karma (hence its name storehouse). In later formulations in Zen, this layer is to be penetrated in much the manner outlined by Maha Bua to release the true No-Mind (Kapleau 1965, p. 361; Thich Nhat Hahn 1974, p. 85). No matter how attractive the bhavanga and alayavijnana, they, like Maha Bua's radiant mind, are not the awakening to pure mind of enlightenment.

What we are left with are formulations that if not identical are certainly broadly parallel. At the depth of mind is a consciousness or nature (bhavanga, alayavijnana, bodhicitta, tathagatagarbha) that is innate, deathless, and marked by radiance. The realization/purification of this nature is enlightenment/Nibbana. Here at the crossroads of high doctrine, the mountain and the forest meet. They also both arrive at an identical intersection, the purity conundrum. How can we already be pure/enlightened? They are compelled to negotiate their way through this juncture of apparent contradiction by similar maneuvers, again indicating their parallel formulations.

Shining Mind in Practice

To this point, we have addressed the shining mind and its correlates as doctrine, but it would be incomplete to understand these teachings as only theory. These categories have tremendous guiding and orienting power. The shining or pure mind is cardinal to the soteriological project and the objective of the Forest Movement's activity. The presumption of these realities and their attainability animates the entire project. Instructions on practice and admonitions to strive onward often invoked the pure mind or heart and its unmatched value. Ajahn Mun lectures his disciple Ajahn Khao Analayo during intensive forest practice:

> This is the ground of the citta, the ground of the Dhamma, the ground of your faith in Dhamma and the ground of the Path, Fruition and Nibbāna—all of them are just there. You must be confident and resolute in your striving if you want to transcend dukkha. You have got to make the effort just there, in the heart. (Maha Bua 2006, p. 59)

The heart/mind is the supreme goal, intimately available, and the very field of practice. However, the shining heart/mind is more. Its degree of realization is perceptible especially to the realized master. Forest accounts not infrequently contain observations about the luminosity of a monk's appearance. The shining mind is, therefore, evidence of the quality of practice, and is in secular terms a diagnostic indicator. Again, Ajahn Mun asks Ajahn Khao Analayo:

> How is your meditation practice going now? Last night your citta was much brighter than it has been at any time since you came to stay with me. (p. 58)

Mun could ascertain this because as he states, "I was sending out my citta to see how you were getting on." Whether through the supra-normal powers claimed by Mun or the pedestrian perceptual channels of an ordinary monk or layperson, a practitioner's brightness of mind can be seen. The shining mind, therefore, can be known in patibat, practice, and performs a diagnostic function.

Summary

The pure mind has been identified as a key teaching of the Thai Forest Movement. Its characteristics were found to be sentience, luminosity, purity, innateness, natural ordinariness, and release. The pure mind is linked by the Forest Movement to two other key notions: unsupported consciousness and Nibbana. Together, these teachings distinguish the Forest Tradition from mainstream Theravada teachings. The promulgation of this doctrine was traced through the generations of forest teachers beginning with Ajahn Mun, the autobiographical expositions and the explicit theory of Maha Bua, and finally to the third generation of Western monks with their more extensive exposition and defense of what remains a controversial doctrine. There is rejection of the pure mind doctrine not only by scholar monk circles but also among some Theravada lay students who see it as a heretical self/eternalism teaching.

Scriptural support for these ideas was located in both the Sutta Pitaka and the Abhidhamma. The contemporary use by the forest teachers of these sources to buttress their dhamma was noted. The notion of pure mind was linked to Pali Buddhist teaching on pure consciousness, unsupported discernment, Nibbana, and the bhavanga, or underlying continuum consciousness. These classic terms were then found to resemble Mahayana concepts of critical importance to Ch'an/Zen, namely: bodhicitta, tathagatagarbha, and the alayavijnana consciousness. In this area of doctrine, the two schools approach one another, but not through any self-conscious project. The similarity of Ch'an/Zen and the Forest Movement was again demonstrated in their common efforts at resolving what was called the "purity conundrum."

Finally, the relevance of these abstract notions to actual practice was delineated. Pure mind and its realization are the telos of both schools. In action, the pure mind is used to inspire, guide, and even evaluate the seeker. If Ch'an/Zen reserves for itself the badge of "direct pointing" to Mind, then the forest school can claim that it is the sect of direct pursuit of the same goal. Both the mountain and the forest proclaim that the true nature of mind is the zenith of all their hard practice.

References

Breiter, P. (2012). *One monk many masters.* Vancouver: Parami Press. Kindle edition.

Bua, M. (1987). *Straight from the heart.* Udon Thani: Forest Dhamma Books.

Chah, A. (2002). *Food for the heart.* Boston, MA: Wisdom.

Chah, A. (2004). *Teaching of Ajahn Chah.* Ubon: Wat Pah Pong.

Chang, G. C. C. (1959). *The practice of Zen.* New York: Harper and Row Publishers.

Cleary, T. (1997b). *Teachings of Zen.* Boston, MA: Shambhala.

Cook, F. (1996). *The transmission of the light.* Somerville, MA: Wisdom Press.

Dhamma Wheel. (2009). Retrieved April 4, 2015, from http://www.dhammawheel.com/viewtopic.php?t=1000

Fuang Jotiko, A. (1999). *Awareness itself.* Valley Center, CA: Metta Forest Monastery.

Harvey, P. (1995). *The selfless mind.* London: Curzon Press.

Jayasaro, A. (2008). *Ajahn Chah's biography by Ajahn Jayasaro by the noble path.* Retrieved January 8, 2015, from https://www.youtube.com/watch#1-43.

Johansson, R. (1969). *The psychology nirvana.* London: George Allen and Unwin LTD.

Kapleau, P. (1965). *The three pillars of Zen.* New York: Random House.

Lee, D. (1994). *The craft of the heart.* Bangkok: Natikul Press.

MahaBua(1976). *TheVenerablePhraAcariyaMun*, SamudraSakornWatPrajayarangsi.

Maha Bua, B. (2006). *Venerable Ajaan Khao Analayo.* Udon Thani: Forest Dhamma Books.

Mendis, N. K. G. (1993). *The questions of King Milinda.* Kandy, Sri Lanka: Buddhist Publication Society.

Mun, B. (1995a). *Ballad of the five khandas* (Thanissaro, Trans.). Retrieved September 3, 2015, from http://www.accesstoinsight.org/lib/thai/mun/ballad.html

Mun, B. (1995b). *A heart released* (Thanissaro, Trans.). Retrieved September 3, 2015, from http://www.accesstoinsight/html/lib/thai/mun/released

Pasanno, A., & Amaro, A. (2009). *The Island.* Redwood Valley, CA: Abhayagiri Monastic Foundation

Sujato, A. (no date). Retrieved September 8, 2015, from https://sujato.wordpress.com

Thanissaro, B. (1993). *The mind like fire unbounded.* Barre, MA: Dhamma Dana Publications.

Thate, A. (1993). *The autobiography of a forest monk.* Wat Hin Mark Peng: Amarin Printing.

Thate, A. (2007). *Only the world ends.* Bangkok: Pattanasuksa.

Thich Nhat, H. (1974). *Zen keys.* Garden City, NY: Anchor Press.

Waddell, N. (2000). *The unborn.* New York: North Point Press.

Walpola, R. (1978). *What the Buddha taught.* New York: Grove Press.

Watson, B. (1993). *The Zen teachings of Master Lin-chi.* Boston, MA: Shambhala.

CHAPTER 8

View, Emptiness, Nowness, and Skillful Means

Introduction

Pure mind is a pivotal idea for the Thai Forest Movement, but there are additional, if more subtle, shifts in the teaching. The notion of pure mind placed mind in the center of forest teaching and practice. While not as explicit as a singular concept like pure mind, there are other reorientations and reinterpretations that have significant implications for how a movement understands the place of doctrine, its overview of path, and the relationship between path and goal. How right view, a primary Buddhist doctrine, is held by Ch'an/Zen and the Forest Tradition will be investigated. Next, three inter-related notions will be highlighted: emptiness, nowness, and wisdom/skillful means. In each case, the emphasis and interpretation give a new cast to not only doctrine but also to practice. This chapter will examine these doctrinal tendencies in both Ch'an/Zen and the Forest Teachings beginning with their understanding of right view.

View from the Mountain and Forest.

The category of view is very explicit in Buddhism. Right view, *samma ditthi* (Pali) or *dristi* (Skt.), is the initial step of the eightfold path. Clearly, right view is foundational and direction giving in both Theravada and Mahayana Buddhism. By one reading to proceed on the path, we need to get our view correct. Again by one reading, this means holding the

A.R. Lopez, *Buddhist Revivalist Movements*,
DOI 10.1057/978-1-137-54086-7_8

correct interpretation of reality which of course is aligned with the core Buddhist concepts such as the Four Noble Truths and the Three Marks of Existence. Yet a closer look reveals far more complexity. By examining working definition of right view in Ch'an/Zen and the Forest Tradition, their fundamental stance on the status of doctrine, the nature of teaching, and their perspective on the soteriological project is revealed. Right view in the narrow meaning of the term, right view in the eightfold path, and in the expanded sense of the term, the overall orientation of the school, will be explored.

Our inquiry into right view will do the following: (a) determine the classical Theravada point of view, (b) identify alternate versions within the early Pali texts, (c) outline the Forest Movement's working idea of right view, and finally (d) note right view in Mahayana Zen tradition and compare it with their forest brothers. The last point will demonstrate that Far East Asian Ch'an/Zen and the Thai tradition hold similar notions of right view in line with their pragmatic-soteriological or psychological pragmatic commitments. Differences in denomination give way to a shared inner dynamic.

Buddhist teaching offers alternate ways of holding (and discarding) Dhamma. There is no more vivid illustration of Dhamma and its truth status than the Buddha's famous Parable of the Raft. The Buddha asks "O' monks what should one do with this useful raft upon reaching the other shore?" He offers various options. Should one put it on one's head or over the shoulder or just leave it beached or submerged? The monks are told, "get rid even of right mental objects, all the more of wrong ones" (Nanananda 1974, p. 39). Perhaps no other passage so forcibly conveys the Buddha's almost ruthless pragmatism. Yet the teaching offers a complex and nuanced message. The raft has been of great value in crossing the great flood to the other shore. The raft is "useful." Not all constructions might have fared as well. There is still a matter of materials and design that may be vital. Along with the function of carrying us to the opposite shore, the Buddha points out that the material chosen were based on what was available. Presumably, this means twenty-first century Buddhist might find different twigs and twine for their raft! Of course, this is not an anything-goes attitude. Presumably, some materials work better than others.

In the predominant Theravada formulation, right view stands in opposition to wrong view. Walking the path entails correcting wrong views and coming to hold the proper versions of the Buddha's teaching and an understanding of reality as it truly is. This transition involves the correct understanding of the propositions of the Buddha's Dhamma. Without this corrective, we are proceeding on the basis of incorrect ideas and our journey is misdirected from the start. This version of right view is delivered in varying levels of sophistication from learned commentary to the discourses of village monks to local children. Talks range from an explanation of basic Buddhist concepts to the inclusion of homey admonitions to love the King and obey one's parents. This is a straightforward version that can be called the right vs. wrong view.

The right vs. wrong model has several characteristics. First, it is oppositional. Right and wrong are two mutually exclusive and opposing categories. The embrace of one and the rejection of the other is the journey into correctness. Second, views right or wrong are about the content of propositions. View means either right or wrong knowledge of an idea or a doctrine. To know that all conditioned things are impermanent is correct view. To believe that there is a permanent phenomenon in the conditioned world is wrong view. The right/wrong model finds its textual basis in the *Brahajala Sutta* with its exhaustive list of 62 improper views. Commentaries elaborating the right vs. wrong model tend to rely on this Sutta. View concerns content. The right vs. wrong view model can also be termed the content or proposition model.

While this is likely to be the presentation in Buddhist catechisms and in the dhamma hall, there is another version to be found in the early texts and, as we shall find, in the forest and mountain. A reading of the suttas and the Abhidhamma suggests, according to a number of scholars, an alternate version of right/wrong view. In the *Sutta-nipata* texts, especially the *Atthakavagga* and to a lesser degree the *Parayanavagga*, a presentation of right view along very different lines is expounded (Fuller 1995, p. 147). Early Abhidhamma treatise such as the *Dhammasangani* also frames the question of right view in terms different from the conventional Theravada teaching (Gethin 1997, p. 211). The key difference between the right vs. wrong model and the understanding of these early texts turns on the concept of *ditthi-raga*, the lust for views. The axis of concern has shifted from right/wrong content to *how* views are held. Psychology has replaced philosophy or perhaps more accurately dogma. Equally, if not more significant than what you believe, is your attachment to views. The presence or absence of clinging lust for views dramatically shapes action. Therefore, it is not only psychology in an inner mode but an existential outer mode of action in the world that is at issue in the right view discussion. This view of views can be called the pragmatic/attachment model.

The Buddha in a famous simile warns against a "thicket of view" that can especially ensnare the monk/nun. The consequence noted by Nanananda Bhikkhu is that the Buddha's reasoning is that the problem is our tendency to be "beguiled" by thought (1974). The Buddha's argument against dogmatism opens up thorny and complex considerations. Going further, the *Atthakavagga Sutta* contains the radical assertion that the brahmin "does not follow views." The pure ones have "no association with knowledge" (Fuller, p. 147). Here, the assertion is that the problem is not merely lust but that all view themselves are the difficulty. Views it seems always entail a hidden lust. Are there in effect no right views? Are all views being biases are to be eschewed? Fuller has dubbed this the "no-views" position.

There are then a variety of ideas even in the Theravada world as to what is right view. Three models have been noted: the right vs. wrong, content/proposition model, the attachment/pragmatic model, and an extreme no-views view. The last two positions identify ditthi-raga, the lust for views, as the cardinal concern, but disagree as to how inherent it is to any degree of identification with a view. It has been called the strong and weak position on holding views. While these distinctions offer conceptual clarification, their operation in the field is a more mixed affair. An example can be drawn from a contemporary Zen master, Maezumi Roshi of Zen Center of LA. One of his monks reported that at meetings he typically refrained from taking a position and then suddenly would declare a decisive point of view on the matter. Hold no view, act decisively seemed to be the teaching. The guiding principle was pragmatic-situational rather than doctrinal. Are correct views, even concerning mundane issues, to be adhered to or are views to be transcended and used skillfully as tools, if at all? The functional understanding of correct view may take different forms in action that cross the boundaries of any philosophical model of correct view.

The Thai Forest ajahans and the sifus and roshis of Ch'an/Zen exhibit a particular slant on the question of right view and doctrine itself. The issue is not merely abstruse theory, but goes to the way that doctrine is carried and transmitted. The very tone and weight of doctrine and how to employ or not employ it is being addressed. The right vs. wrong understanding imposes a litmus test around content, while what we have called the attachment model conveys a different standard of evaluation. The Thai Forest Movement and the Ch'an/Zen School, while not being completely identical, display a similar tendency on the issue of right view. As we would expect their shared concern with the soteriological project will determine their approach to views and doctrine. It is to the actual terrain of the forest and mountains that we now turn.

Ch'an/Zen: No Buddha, No Dharma, No View

The history of Ch'an/Zen is the recurrent dialectical movement of deconstruction and the transcendence of prior doctrine. A persistent teaching of Ch'an/Zen is that there is no Dharma to be held to. These masters are not interested in setting up an alternate theory, which of course would only contradict their own assertion, rather they are insistent that reliance on word-based Dharma is to fall into a pit of delusion. Throughout the centuries, outstanding teachers of the Ch'an/Zen lineages voice this attitude. Ch'an begins of course with Bodhidharma's "outside the scriptures" pointing to Mind or ultimate reality. Whether this pronouncement is history or mythology, it is the Ch'an narrative. There is a knowing that does not rely on texts, and hence written Dharma is a divergence. While more than willing to cite texts in support of its teaching, as evidenced by the early manuscripts of Tung-shan caves and to absorb the formulations of Hua Yen masters and other Chinese schools when suitable, Ch'an /Zen maintains an iconoclastic face.

Even more explicit than Bodhidharma's rather enigmatic declarations is a single line from Sengstan, the Third Ancestor in his *Faith Mind* poem. "Don't seek for the truth, simply cease to cherish opinions" (Clarke 1973, p. 12). The advice sets the perspective of Ch'an/Zen on views and doctrine that have continued for millennia. A tenth century master, Fa-yen, reiterates Sengstan, "I advise you, don't seek reality, just stop views" (Cleary 1997b, p. 45). Rather than promulgating doctrine, Ch'an seeks to strip doctrine away. The orientation is not merely philosophical but is a principle of practice. Give up cherishing view, raga-ditthi, is the way. The downgrading of a conceptual fixed Dharma continues through the teachings of Huineng, the last of the early Ancestors. "Wisdom can only be found by observing the mind, why waste time seeking metaphysical ideas?" (Ferguson 2000, p. 67). Ku-shan sums up Zen's aversion to any and all concepts when he warns "Concepts act as robbers" (Cleary 1997b, p. 42). Prajna, non-conceptual knowing, is of a different order.

If metaphysical ideas are rejected as a way to wisdom and sutra reading discouraged, then what is the status of dharma teachings? What are we to make of right views? Pai-chang, a Tang Dynasty master, puts it unequivocally, "True words are false words when they give rise to views" (Cleary

1997b, p. 14). Here, views as such are seen as obstacles to seeing the truth. Pai-chang quotes a fellow master as declaring that both inward view, various metaphysical positions, and outward views about practices are "both mistaken" (p. 15). Lung-ya who followed Pai-chang by several centuries sums up the Zen position on doctrine. "There has never been any doctrine to give people, just transmission of various expedient techniques" (p. 36). Lung-ya's seems to say there are teaching for the moment but no doctrines, and that the teaching themselves are only skillful means to realize the way. We are hearing the Zen aversion to any reification of views, and the dissolving of all dharma propositions into skillful communications justified by their effectiveness in the moment. Pai-chang clearly states the principle by which all assertions are to be evaluated. "True words are those that actually cure sickness." He goes on to assert, "False words are true words when they cut off the delusions of sentient beings" (Cleary 1997b p. 14). Delusion cutting is the litmus test, so much for formal truth propositions.

Ch'an/Zen is clearly in the camp of no-views theory. Right and wrong have to do with the consequences for ending suffering and attaining the way. In the teachings of Hsuan-shan, all of Buddhism is subordinated to this truth standard. "There is nothing to Buddhism," Hsuan-sha admits, "it can enliven people, and it can kill people too." Therefore, it is "Seeing essential nature and becoming enlightened," that matters (Cleary 1997b, p. 39). His words honor and return us to Bodhidharma. Such pronouncements, however, are not merely the utterances of the ancients, as significant as that would be in identifying Zen's self-branding, but can also be found from more contemporary sources.

The researcher heard the following personal story from Sasaki Roshi, a Rinzai master teaching in the West. In his small village, his birthday was recorded on the Buddha's birthday. Being poor and pious, he ordained as a novice and was sent to the far north, Hokkaido Island, where with two other novices he was ushered in to see the Abbot of his new home. "I have a question for you, When was the Buddha born?" The first boy said, "He was born in North India 2500 years ago." Very good you know history, commented the Roshi. The second boy answered, grandly "The Buddha was never born and never died." Very good you know philosophy. Next, the young Sasaki answered. "He was born when I was." Practically levitating off his cushion, the Abbot stared at him fixedly. "Have the other two study the scriptures. Teach this one meditation."

The story, told with charm and humor, tells us how the Zen tradition regards word-based dharma. Sasaki's tale is told as more than entertainment. The audience is being taught what practice is, namely meditation, and also what is the Zen's opinion of ordinary knowledge. He is branding his sect. The story is a self-presentation and definition of Zen. Knowledge, even of Dharma, is nice, but non-rational intuitive truth is essential. The scriptures are for those of pedestrian understanding. Meditation is the practice of the spiritually adroit. Compare the above with the following from Ajahn Chah, "So the Buddha was not enlightened in India. In fact he was never enlightened, was never born, and never died. The timeless Buddha is our true home, our abiding place" (Chah 1985, p. 179). Such a view undermines any right view based on propositions. Chah speaks from what Zen calls the level of emptiness.

Another event featuring Sasaki communicates an even more radical handling of Dharma. A disciple of Sasaki's who was committed to the idea of a Supreme Being reports that Sasaki would ruthlessly dismiss this notion. "There is no God!" Sasaki would thunder. As the distraught student was exiting the interview room, the old Roshi would whisper, "He's always with you." If right view is correct propositions, then this is madness. If however right view is non-attachment and if Dharma is skillful tactics to cut loose entangled students from a thicket of views, then Sasaki's contradictory utterances are the Dharma of good medicine. A small volume of his teachings repeats this self-contradicting Dharma. Entitled *Buddha is the Center of Gravity* (1974) on the front cover, the back cover declares Buddha is Not the Center of Gravity. Sasaki leaves the reader standing in space with no handhold.

What is notable is that Sasaki is a twentieth century figure who in his hundred plus years bridged into the twenty-first century. This is no T'ang Dynasty mythical figure cloaked in fanciful tales. Whatever the historical truth of classic Ch'an/Zen episodes, they define the movement and provide a working template for contemporary Zen teaching. Certainly, not all Zen teachers treat Dharma so radically, but Sasaki is far from singular, especially among those masters who are viewed as stellar representatives of the tradition. For instance, Sheng Yen, an internationally recognized Ch'an master, when pestered about pure mind shot back, "There is no pure mind ... only the absence of vexations." Once again, a fixed Dharma is eschewed in favor of expedient communication.

In sum, Ch'an/Zen exhibits an aversion to views and dharma propositions, at times seeing them as inherent barriers to awakening. The path is shedding views rather than any exchange of wrong view for right views. Even when views are granted truth value, they are minimized, and if cherished are seen as counter-productive. Zen undoubtedly ascribes to the attachment model of right views, sometimes assuming the extreme no-views position. The Ch'an/Zen stance on views, however, may extend beyond the attachment model. The rejections of all views not only if they are held with attachment assert that their very conceptual nature is ignorance, avijja. The insistence on non-conceptual knowing means that views are hindrances not merely because of the way they are gripped, but also because of the implement itself. Views are ideas and can never accede to ultimate truth, more likely they occlude it.

Views from the Forest

The Forest teacher's understanding of views is implied in their teaching tactics and in spontaneous comments. However, their literature is not without fuller reasoned expositions. Thanissaro Bhikkhu, a Western student of Fuang Jotiko in the Dhammadaro Lee sub-lineage, indirectly speaks to the issue in his essay on not-Self (2015). Thanissaro reviews the Buddha's position on this central Buddhist teaching and then concludes that while the Buddha does not confirm a self, the thrust of his approach is not to set up an alternative no-self doctrine either. From Thanissaro's vantage point, establishing propositions is not the Buddha's preferences. Rather than a new meta-psychology of no-self, Thanissaro finds a strategy to determine not-self. Proposition is replaced by a practice strategy that does not assert a universal truth, no-self, but investigates the status of a here-and-now candidate for selfhood. The question is about the arising phenomena present in consciousness and whether it constitutes a genuine self. What may have been taken as "myself" is subject to testing. A meta-psychology of no-self is not established, rather a particular is found to be not-self. Investigation replaces theoretical axiom.

At the level of practice, Ajahn Sumedho, Chah's preeminent Western disciple, urges at the level of practice a corresponding approach to Thanissaro's not-self strategy. He discourages his student-retreatants from applying concepts such as no-self to their experience. Rather than interpreting their reactions in the light of Buddhist philosophy, he encourages a direct awareness of the mind, of experience as it is (2014, p. 249). Ajahn Sumedho's teaching will be further explored below, but we can already note the strategic/tactical perspective he shares with Thanissaro. Neither are advocates of Buddhism as propositions into which to fit practice/experience which can be a consequence of the right/wrong model of views. As Sumedho confides in his disciples what attracted him to Buddhism was that one did not have to believe in anything. Neither Thanissaro nor Sumedho dismisses Buddhist doctrine, but their handling of it is pragmatic. For Thanissaro, it is rejection of the doctrine, no-self, in favor of a strategy of not-self. For Sumedho, it is warning against the use of doctrine as an evasion of investigation of what is. Wisdom arises from investigation not dhammic propositions.

Both of these monks are Westerners who no doubt entered Buddhism as free thinkers. Nevertheless, they are not alone in their proclivities. In fact, it can be argued that they entered the Forest Movement because they intuited a sympathetic attitude in their teacher's non-dogmatic style. Earlier examples of teaching tactics can be examined for their implied attitude toward views. Ajahn Chah is famed for his admonition, "Don't be a Buddha. Don't be an arahant." The command was taken as so vital to his teaching that it was inscribed on a board and then posted at Wat Nanachat. What is sotapanna, the first taste of Nibbana? Fish sauce, *naam plah*, says Chah. After lecturing on Original Mind, he denies its existence when questioned by a disciple. Are we dealing with a con artist, or a disturbed personality, or a sophomoric prankster? More likely, we are dealing with a teacher who like the Zen masters holds that Buddhism can be either medicine or poison. When the student has swallowed poison, it is compassionate to make him/her regurgitate. Like his Far Eastern colleagues, he is suspicious if not allergic to truth propositions.

Chah's refusal to be pinned down by dhamma concepts indicates that right view is not simply getting the correct answer. This has wide ranging implications. The student is re-focused on the training and her/his raw responses. Sumedho is not interested in your non-self but in the anger that arose in your mind upon stubbing your foot (Sumedho 2014). Chah's teaching insists on the primacy of the monk's training not his intellectual erudition. If the truth is not to be found in language-based propositions, then where is it located? The shared orientation of Ch'an and the Forest monks is that it is in skillful communication and in practice. This common focus of the two schools, however, reflects a working understanding of right view as non-attachment from views.

The pith statement on Chah's view of view is given in a title to one of his short discourses, "Rules are tools" (1985, p. 114). This is a succinct statement of the pragmatic in the Forest Traditions pragmatic-soteriological project. The reification of the pure and the right is seen as a mistake. "We must not cling even to this purity. We must go beyond all duality, all concepts, all bad, all good, all purity, and all impurity" (p. 179). Sumedho, Chah senior disciple, is explicit, "views that are clung to are suffering" (Brieter 2012, loc. 1170).

Emptiness: North and South?

Perhaps no concept is as emblematic of Mahayana Buddhism as emptiness, sunyata. The category is not only a narrow, denotative category, but one that permeates the Mahayana worldview. The root definition stated explicitly by the paramount Mahayana philosopher, Nagarjuna, says that all conditioned phenomena are interdependent and therefore empty of a permanent autonomous self. Certainly, this is a contention acceptable to the Theravada. Sunyata, however, attains an eminence in Mahayana thought that it never achieves in Theravada. We could say that sunyata as a "mark" of conditionality eclipses but does not contradict the Theravada focus on suffering, impermanence, and non-self. The world is empty and, for the Mahayana, this is good news.

Mainstream Theravada discourse gives no such prominence to emptiness, sunna (Pali). For instance, the massive tome ***Buddhadhamma*** written by contemporary Thai scholar monk, P.A. Payutto, dedicates individual volumes to topics such as the three marks of existence, kamma, co-dependent origination, yet sunna receives no such treatment (Payutto 1995). However, the living Theravada, as the Forest Movement has been called, has addressed emptiness and taken pains to distinguish their understanding from their Mahayana cousins. Through the notion of emptiness, a dialogue between the Thai Forest Tradition, the Kammatthana, and Mahayana traditions especially Zen is played out. Two points are notable. First are the differences in Theravada and Mahayana "emptiness" insisted upon by the forest monks. Second is the obvious fact that the subject of emptiness and its comparison with Zen and other Mahayana schools attracts such attention and is a topic of dialogue between the two schools. Emptiness is both a point of contact and a boundary.

Western monks are often the explicit intermediaries in the Theravada/Mahayana and the Forest/Mountain dialogue. Thanissaro Bhikkhu has authored several pieces clarifying the Theravada position as well as posting, translating, and commenting on relevant portions of the Pali Canon where the Buddha takes up the theme of emptiness. These works can be found on the internet (http://accesstoinsight.com). Such visibility speaks obviously to our second point. Discussing emptiness has a level of interest among forest monks that is not found elsewhere in the Theravada world. The content of Thanissaro's work addresses our first point: the differences in meaning of the term between the two schools. Briefly, Thanissaro identifies the Mahayana usage as signifying the absence of inherent existence of phenomena both external and internal. The Mahayana offers a radical critique of ontology, being nor non-being, of reality. In contrast, the Theravada view is more specific and perhaps prosaic, but also more practice relevant. After a review of Canonical sources, Thanissaro concludes:

> Emptiness as lack of inherent existence has very little to do with what the Buddha himself said about emptiness. His teachings on emptiness ... deal directly with actions and their results, with issues of pleasure and pain. (Thanissaro 2010)

Therefore, any implication that emptiness represents an ultimate cosmic principle is canceled in favor of the absence of a particular quality or factor that plays out in practice and the development of equanimity. His investigation finds emptiness as an aspect of meditation, the senses, and finally concentration which culminates in its relationship to wisdom. Thanissaro is insistent that this has nothing to do with a denial of existence (2010). Two other Western monks, Amaro and Pasanno, hold a similar position. They observe that the term appears as an adjective more than as a noun in the Pali texts. A phenomena or process is empty rather than speaking emptiness as does the Mahayana. There is no mystical emptiness, yet they do acknowledge that phrases such as "welling in emptiness" which the Buddha said he often did, and the liberation through emptiness which is noted in the Abhidhamma leaves a different impression. Furthermore, the Buddha speaks of the importance of his teachings on emptiness. With regard to discourses on emptiness, "be eager to listen to them," he advises his monks.

Nevertheless, conventional Theravada, unlike the Kammatthana tradition, seems less than eager to lend an ear. A number of reasons are advanced for this predilection, including those proposed by the forest monks. The term is handled as a meditative insight as opposed to a philosophical understanding. While the differences in emphasis between Theravada and Mahayana on this point and the distinctions on the meanings drawn by the intellectual wing of the Forest Movement may be valid, the reader is left to wonder why the subject has attracted such careful review. Could there be a confluence of practice-based experience between the two schools that call out an interest in emptiness by the Forest Tradition? One is reminded of Ajahn Chah's responses to Mahayana and Zen texts presented to him by a Western disciple. He would largely agree with the ideas but insist on it being grounded in practice (Brieter 2004, loc. 1571). Is it possible that the boundary between Theravada "empty" and Mahayana "emptiness" may not be as firm as asserted?

Ch'an/Zen use of the term sunyata at times takes a form that leaves it open to criticism not only by forest teachers but by its own tradition. Zen does prefer the noun to the adjective, and this may drift into an interpretation of sunyata as a source-womb rather than a quality. However, such a tendency is to be seen as an erroneous reification. Emptiness too is empty, Nagarjuna reminds us. That form and emptiness are inseparable co-existents is the recurrent mantra of the Prajnaparamita Sutras. The postulation of a separate realm of emptiness is denied. Nevertheless, the Theravada remains understandingly suspicious and uncomfortable with Zen's more lavish usage of the term.

However, the very citations by Pasanno and Amaro (2009) raise questions about their conclusion that Mahayana emptiness and Theravada empty are distinct. After establishing the differences between the two Buddhist denominations on the issue of emptiness, the authors move from philosophy and psychology to poetry. The imagery of the two schools mirrors one another. In their concluding section on emptiness, Pasanno and Amaro quote from the Pali Canon. The Awakened One teaches:

> Form is like a lump of foam
> feeling a water bubble
> perception is just a mirage
> volitions like a plantain. (p. 209)

The Diamond-cutter Sutra, so revered in the Mahayana, concludes with this verse:

> As a lamp, a cataract, a star in space
> an illusion, a dewdrop, a bubble
> a dream, a cloud, a flash of lightning
> view all created things like this.
> (Red Pine 2001, p. 27)

There is a difference between de-frocking the khandas of their pretense to a self and the Mahayana extension of non-self to "all created things." Nevertheless, while respecting the careful distinctions drawn by scholars on the use of "emptiness," one wonders why the scriptures expounding emptiness culminate in astoundingly similar imagery. Sunna may not be sunyata, but they both lead to a world in which hard reality has become transparent. Philosophical analysis maintains the coherence of thought, but poetry conveys the texture of experience. The connotative meanings mirror each other. The shared similes invite speculation on a shared experience of emptiness occurring in both the forests and the mountains. Denotative meanings may differ, but teachings that give prominence to sunna/sunyata approach one another in felt texture.

Again, we can ask do the differences in language occlude an identity. How many adjectives add up to a noun? If a traveler to Vermont reports a green flower, a green tree, and a green mountain, and another visitor reacts to the *greenery* of Vermont, are they in different states? Are monks who work with the empty senses, empty concentration, and empty meditation, and who like the Pali Buddha dwell in emptiness in a state different from their Zen brothers? We may never know, but the teachings play off each other and are a place of proximate meaning between the forest and the mountain.

Interestingly, the forest teachings on emptiness also bring the school closer to other Thai lineages that are independent of mainstream Theravada. Certainly, the most significant treatment of the theme of emptiness by a Thai Buddhist is Buddhadasa's *Heartwood of the Bodhi Tree* (1994) which is a full-length exposition of sunna in the Pali Canon. Buddhadasa, of course, is the great rationalizer of Buddhism who, like the Forest Movement, re-focused Buddhism on liberation. Another instance is Luang por Teean, who as a layman developed an innovative meditative technique, and who has a disciple whose thought openly embraces the Mahayana idea of emptiness. In *Know Not a Thing* (1997), Khemananda unabashedly borrows from Zen. Possibly, the movement away from Theravada orthodoxy and its analytical realism permits a less solid, more expansive, view of the world and hence a revived interest in emptiness. In conclusion, with regard to the doctrine of emptiness (or empty), the Pali and Sanskrit in the end seem to agree that the world is less like a brick and more like a bubble.

Nowness and the Way

Aside from the heightened concern with sunna/sunyata, the Forest Tradition takes other doctrinal positions that distinguish it from conventional Theravada teachings. These shifts in overall perspective are less evident than the pure mind doctrine, but perhaps as far reaching. It would be an exaggeration to imply that all forest teachers are involved in these shifts. In general, it seems that the foreign monks in Ajahn Chah's lineage are the most explicit conceptualizers and conveyors of these adjusted views. However, Thai monks are by no means entirely absent from these revisions. For instance, Ajahn Toon in his *Entering the Stream of Dhamma* (n.d.) is unequivocal in his promotion of wisdom, panna, as the factor of first rank on the Buddhist Path over the more common insistence on sila as foundational. The preeminence of wisdom, as will be shown, makes other dhamma claims contingent and subordinate.

Perhaps the most intriguing reorientation is the appeal to "nowness." Certainly, all Buddhist schools acknowledge the importance of present-centered consciousness. Meditation is the cultivation of a mind dwelling in the present. The Buddha urges the letting go of both past memories and future fantasies. Nevertheless, the now is somewhat mitigated in the Theravada given the elaborate teachings on stages of the path and the cultivation of favorable and the elimination of unfavorable mental traits. For instance, conversations with serious vipassana, insight meditation, practitioners in Thailand are typically characterized by a locating of one's self on the path and determining the way station of one's attainment. "I am here, then I will be there." My suggestion to meditators that among the some forest teachers, there is a de-emphasis and perhaps even a rejection of stages attainments brought all most universal discomfort. "We have to know where we are on the path!" was the mode response.

The invocation of the now has critical implications. In contrast to the conventional Theravada and Buddhist framework, the focus on nowness takes the present as more than a practice domain. The way is now. The fruit is now. Nowness is the very field of realization and becomes synonymous with the fruition of the path. The Theravada model is typically construed as a linear path of sequential approximations to awakening. Markers on the trail become critical. Like bases on a baseball diamond, the stations on the path need to be touched in order on the journey to home. Ajahn Sumedho, Chah's senior foreign student, notes this tendency among both his ordained and lay students. What he criticizes is their failure to turn to the now where the conflicts and strivings of the path are immediately resolved. Instead, as he sees it, they take refuge in abstractions, dhamma theory, rather than investigate the dynamics of the experiential present. In a talk entitled *The End of Suffering is Now* (2004b), Sumedho proposes that it is in taking refuge in the Now, which is where pure awareness resides, that suffering ends. This Now, to use his teacher's simile, is the horse we are already riding, but refusing to notice as we stare at the ever receding horizon. He questions as reified statuses both sotapanna and arahantship and even discourages their personalization. "We think that a person is an Arahant or stream-enterer," Sumedho notes, but perhaps these august conditions are non-personal realities present in the Now-moment as universal qualities and fully available if we turn to them. That we ascribe them to this or that monks is "just the way the conditioned mind thinks." In informal conversation with a fellow monk, Sumedho voices the opinion that arahantship is available in the moment as pure awareness. He seems to propose an arahantship that is immediately accessible but overlooked by students blinded by fixed ideas of a path and its fruit.

In a series of talks given at a ten-day retreat in England, we find Sumedho speaking publicly in a manner reflective of his privately expressed vision of the path. He discourages his students from applying the predictable typical Buddhist categories such as "no-self" to their practice in favor of a more persistent examination of their mind states. Of course, mindfulness of mind is a classical domain of investigation recognized in all Buddhist schools. What differs is Sumedho's pointing to an intuitive awareness that is not only the method but the goal. Like the Mahayana and especially Ch'an, Sumedho asserts our inherent wakefulness that pure mind needs to be discovered not constructed. "Awareness is not created" (Sumedho 2004b, p. 90). If so then it is already present. "If we start from ignorance how can we get to wisdom?" (p. 89) he asks in a logic reminiscent of Zen. We are urged to take refuge in intuitive awareness as it transcends the conditioned world and is always present, immanent. The path is no longer the classic Theravada road leading from here to there, but rather the way that contains the goal right here, right now. "It is never apart one where one is—what is the use of going off here and there to practice?" Dogen asks in his *Fuganzazengi* (2002, p. 21). You are already riding the horse you are looking for, Ajahn Chah points out, so look down not ahead.

Sumedho directs us to a non-personal intuitive awareness. That is the work. Awareness is always here. Sumedho's talk, *Now is the End of Suffering*, goes beyond the Buddhist truism that everything happens now including past memories and future fantasizes. Rather the rich adequacy of the moment is being accented. He speaks out vigorously against a mindset that defines oneself as unenlightened, having plans to get enlightened, and generating frustration and despair along the way. This mentality he sees as based on a negative construct of the self. "How do you sustain that one?" It comes and goes. In contrast is awareness—unconstructed and always present. He urges starting from awareness not ignorance. The most radical implication is not that resources are in the now but that the ultimate fruits are right here right now. Now is the knowing, the end of suffering, that is actually happening right now. Practice becomes more a catching it and sustaining it than slogging after it. Sumedho's student Amaro puts it this way, "The end of suffering is not some kind of Armageddon, a cosmic healing at the end of time" (Amaro 2000, p. 13).

Furthermore, by questioning the personalization of liberation by as associating it with a person in the form of an arahant, he implies that it is an error to ascribe to a person what is a non-personal reality available here and now. Arahants are reifications and projections of a truth that can be touched here and now. The pure mind is right here. Chah is definitive on this point when he suggests that when the mind is in pure awareness, one is a "temporary arahant" (Brieter 2012, loc. 1192). Rather than a personal status, arahantship is the nature of now-awareness. The classic Theravada path has been deconstructed. We are in the Zen zone where awakening is immanent, as well as the domain of the *Diamond-cutter Sutra* where the Buddha attains nothing and rejects any personal agents or recipients. We are told monks fainted upon hearing this teaching. If the Chah/ Sumedho's message was received, their monks might have keeled over as well. This is a Dhamma not about a new category but a new orientation which folds the path back on itself. Sumedho introduces not so much a new concept as much as tilting of the teachings at a new angle. His talks telescope the path, normally stretched out to the future horizon, into the present.

Interestingly, Sumedho offers these comments in conjunction with observations on the recent draw of Zen and Dzogchen (a Vajrayana practice) for Theravada practitioners in the West. What Sumedho hears from these students is their frustration with piecemeal mindfulness that is always on the road but never arrives home. He refrains from sectarian arguments, but his subsequent comments direct the practitioner into the Now where according to Zen (and Dzogchen) the awake-nature resides. Concepts such as Buddha and Dhamma in which Buddhists take refuge are reframed in experiential terms: awareness and things are they are. Even more radical is his contention that in Now-awareness, there is no suffering. Now is the end of suffering. Now is the place of freedom.

This perspective brings Sumedho perilously proximate to Zen. *Hongaku*, (Jp.), *Ben jue*, (Ch.) is the Zen doctrine of original, innate enlightenment that is shared with other Mahayana schools such as Shingon and Tien-tian, and which first prominently appears in *The Awakening of the Faith in the Mahayana Sutra* which so influenced Ch'an. This doctrine makes sense of the Ch'an/Zen dictum looking into the mind and realizing Buddha. If the enlightened nature is already present, then it is a matter of seeing it. In answer to my question, What is enlightenment to you? Gouyuan Fashi responds, "Seeing your true nature." In understanding the Now to be replete with virtues and the field of realization, Sumedho, perhaps without intending to, has given Theravada a Zen spin. In a talk entitled "Original Universal Purity," Sumedho directs us to "Contemplate purity here and now" (2014, p. 288). The gates to the deathless are open here and now because the deathless is here-and-now.

We should be cautious and acknowledge that Sumedho's views are not necessarily shared by other forest teachers. Those practicing in the Maha Bua sub-lineage tend to be more conservative in their presentation of Dhamma, despite their own teacher's innovations. On the other hand, Sumedho's mentor, Ajahn Chah, seems to pioneer his disciple's teaching when he brusquely tells his monks do not be a Buddha, do not be an arahant. His characterization of the entry stage of awakening as mere flavor enhancer, fish sauce, opens the way to Sumedho's Zen (and Dzogchen) like perspective. While not universal to forest teachers, these innovations are emergent currents within the Thai world that flows toward the rivers of Ch'an.

The contention that awakening is a radical redirection of the mind rather than a building up of favorable factors mirrors Ch'an/Zen. If the much vaunted sudden enlightenment of Zen has any meaning, it is perhaps the insistence that awakening is a sudden shift in perspective in which the Now appears as the complete repository of the sacred. In the Fukanzazengi, Dogen advises us to stop chasing words and traveling to far off lands as the truth lies beneath your feet. "The treasure store will open of itself," he tells us (Cook 1978, p. 98). "This very Earth is the Lotus land of Purity," Hakuin declares in his *Song of Zazen* (Suzuki 1978, p. 152).

The fusion of the immediate moment and the Ultimate imparts startling power to Zen verse. Dogen's enlightenment poem reads:

> Sitting awake
> I hear the One True sound
> black rain
> on Fukawara temple. (Styk 1994, p. xlv)

The vision extends into practice. Sasaki Roshi stands outside his hut, thumping his staff on the wooden floor. This is the same sound President Roosevelt heard. (Remember he was well into his nineties.) Demanding of me "What is this sound?" A Western Ch'an teacher offers the slogan, "Practice for now"(Child 2006). Thich Nhat Hahn (1974), the famed anti-war activist, tells us that peace is in every step not in a distant political triumph.

In such a view, the path and goal telescope into the moment. When Reb Anderson, a successor of famed Soto teacher, Suzuki Roshi, was asked about stages of practice by a visiting Tibetan lama, no stages, he replied. A very advanced practice the lama commented. An American student tells his Zen teacher he wants to deepen his practice. The answer he receives is there is no deeper. The student is being driven uncompromisingly into the present where the body of dharma, the Dharmakaya, resides. Here, Ajahn Sumedho might say, is the end of suffering. Zen agrees.

The Triumph of Wisdom and Skillful Means

The link between Buddhism and wisdom is hardly new. However, wisdom in the view of the Forest School and Ch'an/Zen climbs to a supreme position. For the teachers of the forest and the mountain, all doctrines and its claim to truth are subordinate to wisdom. The wisdom in these schools is of course not a set a dhammic propositions, rather it is panna/prajna. The term is generally translated as wisdom but is perhaps better understood as a non-conceptual knowing or intelligence. Elsewhere, panna may refer to the first three steps on the eightfold path, but the usage that attracts teachers in both schools is panna/prajna as non-conceptual awareness. Huineng could be called the master of prajna given its import in his Platform Sutra. In his Chap. 2 dedicated to prajna, he tells his audience that it "comes from one's essential nature: not from the outside." Prajna is the "essence of mind." The free and unimpeded flow of prajna is non-other than the awakened mind (Red Pine 2006, p. 22).

Huineng's understanding has its rough correlate in the Theravada's vipassana-panna, which is the insight awareness that flashes forth and is the key factor of Awakening. In Ajahn Maha Bua's slim volume, *Wisdom Develops Samadhi* (2005b), the classic sequence of concentration (samadhi) leading to wisdom (panna) is reversed. The direct activation of prajna can lead to calm. Ajahn Toon placed wisdom, panna, before sila, precept. As in the Ch'an/Zen school, the prominence given wisdom in the Forest Tradition changes the status of doctrine and right view. If the goal is liberation and if the unleashing of a non-conceptual knowing is what liberates, then doctrine and views are, at best, preliminary guides and at worst fetters to the true seeing-eye. The insistence on panna/prajna over scholarly or cerebral knowing is what separates these practice lineages from the scholar monks or devotional sects. The logic of their stance consigns view and doctrine to an inferior position and indicates the triumph of wisdom.

The second and related principle that triumphs over views/doctrines as fixed ideas is skillful means. Wisdom is what liberates and skillful means is what efficiently and effectively cultivates or uncovers wisdom. Skillful means, *upaya*, is a term dear to Mahayana schools as the active, engaged activity of the Bodhisattva. In the Pali, *upaya-kausalya* is its equivalent but without the same notoriety. The terms occurs only once in the Canon but with greater frequency in post-Canonical writings (Gombrich 1997, p. 17). The sparse usage in the Theravada only makes the actions to which the terms point all the more noticeable.

Even without a handy label, the Forest School invokes the principle of skillful means. As we saw in their teaching tactics, behaviors that might be considered rude, bizarre, and un-monkish are legitimized by the pragmatic imperative. Wisdom serves the soteriological, while skillful means is the pragmatism. Together, they constitute the operating factors of a pragmatic-soteriological project. The triumph of wisdom/skillful means over right/wrong views model is the corollary of the pragmatic-soteriological project. Each of the three concepts, emptiness, newness, and wisdom/skillful means, reframes the overall shape of the path and practice. Emptiness creates a more spacious and fluid view that gives a malleability to practices and supports a more indeterminate path. Tight categories and linearity are softened. Nowness encourages a conflation of path and goal. It is all here and now. Lastly, skillful means subordinates doctrine to the pragmatic, bringing our conversation back to the initial topic of right view and the attachment/pragmatic model.

Knowledge and Emancipation

To gain an overview of what is knowledge in the in the Forest School and Zen, the thought of Jurgen Habermas of the Frankfurt School is relevant. Rather than assume that all knowledge systems are the same merely better or worse, Habermas identifies three separate cognitive interests that characterize knowledge systems (1971, p. 6). The cognitive interest is the objective and type of knowing. What constitutes knowledge? These cognitive interests make for different types of knowing that align with their respective objectives. It is not a matter of right or wrong but of different.

For instance, science seeks efficient prediction. In contrast, there is the understanding, *verstehen*, of a dream, a work of art, a culture. There is, he argues, a third cognitive interest, neither prediction nor understanding but emancipation. As a thoroughly German thinker Habermas locates Marxism and psychoanalysis in this category. However, widening our sample, we discover that there are other projects that share the commitment to transformation. Religious traditions provide not only "meaning" but also the *means* to transform the human condition. Religious systems console, offer guidance, and legitimize power, but they do more. Granting the conservative role of religion which has been amply documented, there is another current to be found in varying degrees in the world religions. From the mystical to the millennial, religious systems have been blue prints for transformation, either at the collective or individual level.

Each of Habermas' modes of knowledge has its own epistemology. For instance, we do not predict dreams, we understand them. Similarly, transformative knowledge systems are not to be evaluated by their predictive efficiency. Prediction of an objective world apart from the observer is the domain of science. The emancipation type has an alternate method, self-reflection. Unlike science or pure understanding, transformative systems demand a self-reflexive stance that emancipates through bringing into awareness the forces of conditioning. Buddhism's affinity with this mode is evident. The Buddha urges listeners to test his teachings (Wallis 2007, p. 24). The truth test is pragmatic: when teachings are correctly applied, are there emancipatory outcomes. Undoubtedly, other religious traditions fit this category. One thinks of Sufism, Advaita Vedanta, Taoist alchemy, and Hellenic mystery schools as transformative projects.

Within the transformative type, there is an action component that alters the entire status of knowledge, and, in the case of Buddhism, Dhamma. How a teaching has truth value and how that value is verified is now based on considerations of action/consequence for both teacher and taught. Fuller in his work on ditthi (2005), view, in the Theravada notes the action component as being a critical difference between the right vs. wrong model with its concern with truth propositions and the attachment model with it focus on how view are held and its implications for action. In the context of Habermas' categories, action or practice means emancipation. Their insistence on the actor's engagement and the overarching concern with outcomes means that the Thai Forest and the Ch'an/Zen Schools conform to Habermas' emancipatory knowledge type. Seen within this model, it is clear that the usual kind of knowledge, even religious knowledge of meaning giving, is replaced by a Dhamma not of propositions but of communications in the immediate service of the soteriological project. Dhamma as statements as Truth-reality give way to a functional dhamma that serves the psychological pragmatism and soteriological-pragmatic project of the forest and mountain monks. In its extreme formulation, it means there is no Dharma apart from liberation; there are only expedient means.

Summary

In this chapter, we have examined broader shifts and tendencies in the teaching of the Thai Forest School and how that has re-positioned it in relationship to Ch'an/Zen. The seminal notion of right view was found to be handled in several ways within Buddhism. Right view is understood as a contrast with wrong views in the conventional understanding, while a second model evaluates view based on attachment and consequences for liberation rather than as dhammic propositions. Teaching is not to establish a correspondence with an objective reality, but is rather a means for transformation justified by its effectiveness. A radical camp in this approach terms all views as limiting. Liberation is holding to no views.

The reorientation of practice and path was explored through three concepts: emptiness, nowness, and skillful means. In each case, they were shown to have been given a new tilt that shifted the Forest Tradition's world view toward the Zen School. A more indeterminate world, a closer relationship between path and goal, and the subordination of dhamma to pragmatic concerns bring the forest closer to the mountains. The knowledge systems of both schools were then located within Habermas' transformative type with its cognitive interest in radical change in the human status. Knowledge in Ch'an/Zen the Kammatthana is a transformational epistemologies in the service of liberation. Again the defining dynamic of the forest and the mountain is found to be their pragmatic-soteriological project which directs how they understand and employ right view and dhamma teachings.

References

Amaro, B. (2000). *Theravada buddhism in a nutshell.* Redwood Valley, CA: Abhayagiri Monastery.

Breiter, P. (2004). *Venerable father.* New York: Paraview.

Breiter, P. (2012). *One monk many masters.* Vancouver: Parami Press. Kindle edition.

Buddhadasa, B. (1994). *Heartwood of the bodhi tree*. Boston, MA: Wisdom.
Chah, A. (1985). *Still forest pool*. Wheaton, IL: Theosophical Publishing House.
Child, S. (2006). *Practice for now*. Retrieved from http://www.westernchanfellowship.org/lib/wcf
Clarke, R. (1973). *Faith in mind*. Buffalo, NY: White Pine.
Cleary, T. (1997b) *Teachings of Zen*, Boston: Shambhala.
Cook, F. (1978). *How to raise an Ox*. Los Angeles, CA: Center Publications.
Dogen, E. (2002). Fukanzazengi. In J. Lori (Ed.), *The art of just sitting* (pp. 21–24). Boston, MA: Wisdom.
Ferguson, A. (2000). *Zen's Chinese heritage*. Boston, MA: Wisdom.
Fuller, P. (1995). *The Notion of Ditthi in Theravada Buddhism*, New York, Routledge Curzon.
Gethin, R. M. L. (1997). Wrong view and right view in the Theravada Abhidhamma. In *Recent researches in Buddhist studies* (pp. 211–229). Colombo.
Gombrich, R. (1997). *How Buddhism began*. Delhi: Munishiram Manoharlal.
Habermas, J. (1971). *Knowledge and human interests*. Canada: Beacon Press.
Khemmananda. (1997). *Know not a thing*. Bangkok: White Lotus Press.
Maha Bua, B. (2005b). *Wisdom develops samadhi*. Udon Thani: Forest Dhamma Books.
Nanananda, B. (1974). *The magic of mind*. Kandy, Sri Lanka: Buddhist Publication Society.
Pasanno, A., & Amaro, A. (2009). *The Island*. Redwood Valley, CA: Abhayagiri Monastic Foundation.
Payutto, P. A. (1995). *Buddhadhamma*. Albany, NY: SUNY Press.
Red Pine. (2001). *Diamond sutra*. New York: Counter Point.
Red Pine. (2006). *Platform sutra*.
Sasaki, J. (1974). *Buddha is the center of gravity*. San Cristobal: Lama Foundation.
Stryk, L. (1994) *Zen Poems of China and Japan*, New York, Grove Press.
Sumedho, A. (2004a). *Intuitive awareness*. Penang: Inward Path.
Sumedho, A. (2004b). *The end of suffering is now*. Retrieved June 6, 2015, from http://www.amaravati.org/audio/the-end-of-suffering-is-now
Sumedho, A. (2014). *Direct realization*. Hertfordshire, UK: Amaravati Monastery.
Suzuki, D. T. (1978). *The Lankavatara sutra*. New York: Grove Press.
Thanissaro, B. (2010). *The integrity of emptiness*. Retrieved September 9, 2015, from http://www.accesstoinsight.org/lib/authors/thanissaro/integrityofemptiness.html
Thanissaro, B. (2015). *No-self or not-self? in Noble Strategy*. Valley Center, CA: Metta Forest Monastery.
Thich Nhat, H. (1974). *Zen keys*. Garden City, NY: Anchor Press.
Toon, K. (n.d.). *Entering the steam of dhamma*. Udon Thani: Thammasat University.
Wallis, G. (2007). *The basic teachings of the Buddha*. New York: Modern Library Paperback.

CHAPTER 9

Meeting of the Twain: Revivalism and Modernity

Introduction

The Thai Forest Movement and twentieth/twenty-first century Zen/Ch'an live in changing times. They find themselves at crossroads not only of East and West but of the traditional and the modern. Far from passive by-standers, they have been shaped but also stepped into and shaped a modern Buddhism. The encounter has been more than narrowly geo-political. Ways of thought and action alien to their traditional forms have been confronted, absorbed, and even pursued. We will explore their meetings with modernism and the similarities and differences in their responses. The ambiguities and contradictions that constitute the tensions of the two schools and modernism itself will be identified.

Both these Buddhist schools have been understood as revivalist in their insistence that their project is the revival/maintenance of the Buddha's essential message and intent. In contrast to the ossified and perhaps even degenerate Buddhism around them, they practice, according to their narrative, a pure dharma/dhamma that was the authentic heritage of the Awakened One. For the Thai Forest monks, this inheritance was both the meditation practice and the scrupulous adherence to the Buddha's code of discipline, Vinaya. Ch'an/Zen, as we have seen, claims a direct mind transmission reaching back to Buddha. With the flourishing of their sect the Buddha Way has been preserved. With this seemingly backward glance, they would predictably be ill fitted to encounter the modern. Yet

A.R. Lopez, *Buddhist Revivalist Movements*,
DOI 10.1057/978-1-137-54086-7_9

in meeting the modern they reveal a complexity in their attitudes and inspiration that has surprisingly projected them into the vanguard of Buddhist modernism. They exhibit a Janus-faced stance that looks both back to a pristine source and forward to a Buddhism suited for the future. The precise relationship between the revivalism of the mountains and the forests and modernism is an intricate dance.

First Encounter

There is perhaps no more surprising than the traditional Thai Forest Movement transitioning into the contemporary. As a specific social movement within the broader tradition of forest monks, it has its beginnings in the early twentieth century. In the year 1900 Thailand was estimated to have a population of one million overwhelmingly concentrated in the central Chao Phraya River valley. This leaves Isan, the geographically largest region of Siam and cradle of the Forest Movement, as a sparsely populated wilderness. Government efforts to recruit settlers included the wholesale relocation of Lao persons to populate the hinterland. In the first decades of Ajahn Sao and Mun's wanderings, modern transportation was unknown. Communication was carried out village to village by official runners who in a land without telegraph service were the common means of conveying news. A trip from Nong Hai in the north of Isan to Ubon in the south took over a week. Spirit worship with a sprinkling of Buddhist magic and simple devotionalism was the religious landscape. One could hardly think of a more unlikely environment for a species of Buddhist modernism to arise that would be successfully transplanted across the world. In less than 100 years, the Forest Tradition would have spawned branches in Australia, Europe, the UK, and North America. Here is a story that is as revealing of the Forest movement's dynamics as it is astounding.

Unlike the Thai Forest Movement, Ch'an/Zen is hardly a discrete phenomenon of a single nation and century. Discussion of Ch'an/Zen's engagement with the modern therefore requires selectivity. Granting the intellectual value of a study of the Western impact on Son and Thien, the Korean and Vietnamese versions of Ch'an respectively, such an effort would make the discussion unwieldy. This is not intended to deny the omnipresent historical fact that three of the four countries hosting the Ch'an/Zen tradition have felt the influence of Western culture and also the impact of American military might and occupation. The effect of Western power on Korean and Vietnamese Buddhism is a significant topic for East Asian religious study requiring its own platform. Therefore, the thrust of our investigation, as it has been, will be limited to Chinese Ch'an and Japanese Zen.

In contrast to the Thai Forest Movement, Ch'an/Zen especially Japanese Zen initially seems to have a more evident exposure to modern influences. Japan's famous pivot to the West in the nineteenth century is a springboard to Zen's journey to the West. More than passively awaiting the avalanche of Western culture, Zen and Japanese intellectuals were engaging western style modernity on their own soil and terms. Even as the West exhibited a fascination with Zen in the twentieth century, this cross-global interest was being anticipated by Japanese intellectuals and Zen modernists who found sympathy with the cultural currents of the Hudson and Thames Rivers. China, the home of Ch'an, may appear at first glance more remote to modernist influences, but we must remember that Sun Yat Sen was a Christian modernizer, and as we shall see, Taiwan, the last redoubt of his Kuomintang movement has become the epicenter of Ch'an Buddhist modernism.

The other agent in the encounter, modernism, requires definition. Modernism has achieved its identity without regard to Buddhism or for that matter Asian culture. Initially modernism was a wholly western concern. The term attained its definition as a contrast to the culture that preceded it in the West. In its narrowest formulation it was a literary and artistic impulse. The definition has been broadened from a purely intellectual/artistic outlook found in high culture to a social consciousness, a view of the world, which characterized an entire culture. Scholarly studies have identified themes that characterize modernism across national boundaries. These build upon and elaborate classical observations such as Max Weber's notion of disenchantment, the striping away of non-rational and pre-scientific thinking. Weber's inheritors extended his insight to an investigation into the unraveling of the sacred canopy of religious meanings which once covered society.

Further reflections on modernism not only broadened the term but incorporated contradictory impulses. Modernism was now more than the steamroller of secular disenchantment riding on Weber's rational and routinized modes of production and administration. Modernism was also the reactions against the modern world; a new set of obsessions that pined for a transcendence or reversal of the everyday modes of contemporary social consciousness. Modernism was now also anti-modern. A key contributor to this discussion is Charles Taylor who in his seminal *Sources of the Self:*

The Making of Modern Identity (1989) tabs Romanticism of eighteenth century Western Europe as a major contributor in the making of the modern self. The limitations of science, the love of nature, and above all the exaltation of the non-rational experience and the inner psychological space from which it issues forth were not only integral to the modern self, but were to become key points of receptivity in the Buddhist/modernity encounter.

McMahan (2008) enumerates three domains that constitute the themes of the Buddhist/modernity interface: (1) The theistic encounter, (2) Scientific naturalizing, (3) Romantic expressiveness (pp. 10–11). By theistic encounter, McMahan indicates the religion-to-religion (or theology to Dhamma) interface. Scientific naturalizing conveys the valuing of rational, conceptually coherent thought and method. Romanticism is seen as a valuing of the subject-self and an effort to re-enchant the world. Already we see modernism by the above definition as containing conflicting directions. Modernity, McMahan observes, is "not homogeneous" (p. 14)

Donald Lopez, like McMahan, is concerned not with modernism in general but with its operation within the Buddhist world. Lopez (2002, p. x) sees contemporary Buddhism as a global sect that flies above regional cultures and is held aloft by "reason, empiricism, science, universalism, individualism, tolerance, freedom and the rejection of religious orthodoxy." The list is concluded with an inclusive stance toward women thereby becoming a laundry list of political correctness and Western liberalism. He also offers a thumbnail synopsis of the modern culture that confronts Buddhism: "the emphasis of the mechanical over the organic, the individual over the group, differentiation over unity, the real over the transcendent, the existential over the metaphysical" (p. 186). As Lopez' study of Buddhism and science indicates, Buddhism's interface with these orientations is complex, selective, and varied over time. We are alerted to multiple levels of analysis: Buddhism's self-representation, the imaginings of new adherents, the actual fit with modernism at the level of social action, and what is shown through a more dispassionate analysis of the ideas of modernism and Buddhism.

Other Buddhologists have explored Buddhism and modernism, especially the Sri Lankan case, which offers several guiding concepts and insights that can inform our study. Gombrich and Obeyesekere (1988) have used of the notion of Protestant Buddhism to characterize the reason friendly Buddhism forged in the anti-missionary crusade and the Sri Lankan struggle for independence. This concept includes themes noted by other thinkers about modernism, such as rationality, individualism, egalitarianism, and privatized spiritual experience (McMahan 2008, p. 7). The Buddhist scholar Hans Bechert's (1984) effort to delineate Buddhist modernism defines it as a revivalist movement that reinterprets Buddhism as a rational doctrine of meditation and a philosophy of life congruent with modern values of equality and democracy. His identification of processes such as de-mythologizing and de-ritualizing are particularly relevant. The designation of modernism as a revivalism corresponds to the definition that has been used to define both the Thai Forest Movement and Ch'an/Zen in this study.

The Background

The story of modernism in contemporary Zen and the Forest Tradition often begins with a memorable face-to-face meeting between famous monks and would-be disciples and interested academicians. Soen Roshi at the World Parliament of Religions in Chicago or monk Sumedho meeting Ajahn Chah at his forest monastery are two such events. However, the background of modernist influence, here meaning Western influence, predates these incidents. Furthermore, in pulling back the curtain to view larger historical processes, the catalytic role of the state in the origins of Thai, Chinese, and Japanese Buddhist modernism is revealed. In each case, the monarchy was a catalyst for the reconfiguration of its nation's Buddhism. In the Thai case, the monarchy promoted its brand of modernism ostensibly to restore true Buddhism but also to promote its own political agenda. In the Chinese and Japanese cases, the Imperial regime unintentionally stimulated a Buddhist reformation through its suspicion of the Sangha. The varied stances of royal authority and the state would play out in the different modernist dynamics of the Ch'an/Zen and Thai Forest Movements.

Thai modernism and Thai Buddhist modernism begin with King Mongkut, Rama IV, of the Bangkok-based Chakri dynasty. The narrative learned by every Thai pupil is that after 27 years in the monkhood Mongkut assumed the throne. As the protector of the Sangha, a responsibility traced back to the Emperor Asoka, Mongkut did a temple tour and returned to the palace aghast. What he found was slack discipline and superstition that bore no resemblance to his dhamma education and practice. Mongkut's reaction, however, represents not only his monk education but his wider exposure and interest in Westerners and their ideas. His

Buddhism like the "Protestant Buddhism" of Sri Lankan reformers was a rational doctrine capable of holding its own in inter-religious debate and at home with the West's Enlightenment.

The direct connection with the Forest Movement follows from Mongkut's bold reform of founding a new nikaya or order within the Sangha that would embody strict discipline and pristine doctrine. The Thammayut nikaya gave organizational form to Mongkut's modernist agenda and was to become the institutional home of the forest wanderers. However, it was not until the reign of his successor Chulalongkorn, the "Great Modernizer," that the Thammayut and its text-abiding monks began to re-shape the Sangha. With his half-brother, a Thammayut monk, installed as the Sangharaj, reform took hold. Under a series of laws and decrees promulgated in the early twentieth century, the Sangha was centralized and routinized, and the Dhamma trimmed into an official catechism.

The simultaneous rise of the contemporary Forest Movement and the bureaucratization of the Siamese Sangha were not a singular movement but interactive tendencies often in tension with one another. One monk's modernism is not another's. For the Bangkok center, modernity was not only rational doctrine but rationalized social organization. For the revivalist forest movement wandering in the margin, Dhamma meant not a catechism but the liberation teachings of the Buddha and the avoidance of the Sangha bureaucracy wherever possible. Where the modernism of the Crown and the forest coincided was in a preference for a logically coherent doctrine and a precept abiding monkhood, along with a shared disdain for local tradition and what they deemed superstition. The modernism of the Great Modernizer and the Forest Monks was only partially aligned. The peripatetic asceticism of the forest monks expressed a more primordial heritage which left the Thammayut bureaucrats ill at ease. Nevertheless, the background to the contemporary Forest Movement includes this broader modernizing trend emanating from the Crown both as a general sensibility and as an organizational support and framework. What defined the Forest Movement was the fusion of Vinaya discipline, a rational dhamma teaching and a fierce ascetic and meditative practice (Jayasaro 2008, p. 10).

Like the Thai case, modernist Buddhism in both China and Japan was stimulated by the state but with a significant difference. Unlike the Siamese Crown, Imperial power in the Far East did not view Buddhism as the preferred ideological agent for building a modern state or sustaining the empire. Buddhism was seen not only as antiquated but as also a foreign doctrine not fit to support either nationalism or emperor worship. While not being subjected to the persecutions that mark earlier Chinese Buddhist history (564, 567, 845, 955 C.E.), Buddhism, and more specifically Ch'an, was not favored at the court. In the early twentieth century, the Qing moved against Buddhist landholdings. The favored Buddhism was Vajrayana, imported from Tibet. This left Ch'an/Chinese Buddhism looking for a way to project itself into the twentieth century. Sheng Yen, an eminent modernist, for example, goes out of way to insist that Ch'an was the Buddhism of the masses while Vajrayana was confined to court circles (2010, p. 48). If Chinese Buddhism was to survive, it needed to respond to the modern. Ironically, being held at arm's length by the Qing emperors provided the space for a positive engagement with Western modernism.

The modernist impulse in Chinese Buddhism found its primary advocate in Taixu, who formulated Buddhism for Human Life which criticized the emphasis on ghosts and funerals that he saw in the late Qing and early Republican periods. Sheng Yen, one of his famous four disciples who came to be known as the Four Dharma Kings, conveys his teacher's sensibility in his autobiography where he expresses distaste for the performance of funeral rites. Taixu claims inspiration from radical Chinese political writers. "I came to see Anarchism and Buddhism as close companions, and as a possible advancement from Democratic Socialism." As a young man, he engages in revolutionary actions against the Qing. Yet ironically, it is with the defeat of the Kuomintang by the Communists and their flight to Taiwan that Taixu's movement takes mature form as "Humanistic Buddhism" (Pitman 2001, p. 182). Our examination of contemporary Ch'an modernism, especially in Taiwan, must acknowledge its roots in Taixu's Humanistic Buddhism. As in the case of Siam and the Thai Forest Movement, the background for modernist developments was not an unmitigated imposition of Western culture but a proactive engagement by Asians reforming their own traditions.

Japan's opening to the West and the Meiji Restoration, like the Chinese experience, resulted in a slighting of Buddhism by the central power. Buddhism was seen as anti-modern and non-Japanese. And yet like their Chinese Buddhist brethren, this became a challenge for Japanese Buddhists to find new forms both symbolic and organizational. The New Buddhist Movement of Japan sought to align Buddhism with the modernization of Japan. Modern intellectual currents were incorporated, and like Taixu's Humanistic Buddhism, social welfare projects were launched.

A significant member of the New Buddhist Movement was none other than Shaku Soen, who would represent Japanese Zen at the World Parliament of Religions (McMahan 2008, p. 64). A later modernist Japanese movement was the Soka Gakkai, which not only has added to the Japanese Buddhism but also has projected itself worldwide (Hughes Seager 2006). Perhaps the most significant feature shared by all three cases, Siam, China, and Japan, was the conviction that their aim of a modern Buddhism could be attained only by returning to a pure, authentic Dharma devoid of popular accretions. Revivalism and modernism, in their view, were not contradictory currents but a single flow of Buddhism simultaneously ancient and new. Revivalist/modernist Buddhism is a back-to-the-future project.

This confluence of revivalist and modernist tributaries is further muddied by the cross-currents contained within modernism itself. The literature offers clusters of overlapping dimensions that define modernism. Some as we have seen are general, abstract categories that are pitched at the intellectual/cultural level (Taylor 1989). Modernism is not only the prevailing secular consciousness but also the reaction to that consciousness. Modernism is not only the secular view but also an effort at re-sacralization, a disenchantment but also a re-enchantment, a qualified respect for science and yet a romantic communion with the trans-rational. Therefore, manifestations of Buddhist modernist hybrids in particular historical/cultural contexts are complex if not apparently contradictory phenomena. In a comparative investigation of the Thai Forest Movement and contemporary Ch'an/Zen, we will examine their reformed practices and actions in the world. Our study will begin with the symbolic level, de-mythologizing, de-ritualizing, attitudes toward science and psychology, and romanticism and the privatization of spiritual life. Then we will move on to engaged Buddhism, modernism in action.

Modernism in the Mountains and Forests: De-Mythologizing

De-mythologizing refers to the expunging or least diminishing of supernatural elements in theory, group narrative, and practice. Myth here is, of course, not the myth of contemporary mythologists who study of myth to mine for universal patterns of experience or to place myth in relation to other social forms. Such study is itself a case of modernism in action. The myth from religious systems is a story typically transcending natural laws that is presented as literal truth and belief in which is often required.

De-mythologizing clears the symbolic space for two often-noted tendencies in Buddhist modernism: sympathy with modern science and psychologizing. An unreconstructed mythical view blocks a peaceful coexistence with science and with the pragmatic self-cultivation fostered in both Zen and the Forest Tradition. De-mythologizing, therefore, is a crucial maneuver that positions each movement not only in its interface with the modern but also in its pursuit of liberation. De-mythologizing is more than good advertising for external consumption; it is the creation of a symbolic space internal to Ch'an/Zen and the Forest Tradition that is suited for liberation through self-cultivation. Practice is pragmatic, rational, and accomplished through self-effort. The gods are sidelined.

The concern with the practice leading to awakening renders much of the metaphysical constructions in Buddhism irrelevant. For instance, the *Triaphum,* a Thai Buddhist document favored by the ruling elites and Buddhist officialdom as a justification of a class society that lays out hierarchical cosmology, is cited in parliamentary debates but goes unreferenced by the forest dwellers (Jackson 1993, pp. 77–79). Likewise, while observed Zen centers did have seminars on sutras that contain fantastic elements, these were treated as dramatic and figurative devices and rarely entered into dharma talks (*teisho*, Jp.) by western teachers. They were marginal and the student was allowed to handle them without intrusion into their modernist worldview.

Both the mountain and the forest schools retain mythical features. Ajahn Mun's hagiography takes the reader into a magical terrain replete with ghosts and giants. Mun famously shared a cave with a monk hating giant who he converts into a dharma disciple. We are in a period closer to Homer's Odyssey than a modernist novel by Proust or Joyce. This interplay with the supernatural finds commonality with Buddha's interactions with gods and *nagas*. Mun's tale aligns with the Buddha and invokes the romantic but does so at the risk of incredulity of the contemporary reader. Recalling the setting of early twentieth century Isan with its wilderness and spirit worship, Mun's account understandably conveys a premodern consciousness. Notably, supernatural elements disappear in later Thai Forest biographies and dhamma talks. Mae Chee Kaew's life is a tour of non-ordinary experience, but much of this is placed in a psychic realms and powers recognized by classic Buddhist meditation manuals. Maha Bua's harsh criticism of Mae Chee Kaew's fascinations puts the entire narrative a sober perspective. De-mythologizing may not entail the complete elimination of myth but only the marginalizing of the supernatural, so that it is no longer vital for the soteriological project. Indeed, this is the case for both Mun and Mae Chee Kaew's narratives.

Recent Ch'an studies have unearthed far more magic than an occasional fox-spirit. Zen it seems was infested with the spirits and gods of its culture. Not only were foxes prevalent but even a figure such as Dogen was far more involved with the gods than Zen's modern self-presentation would have allowed. Dogen's thought has come to us through works of philosophical reflection such as *Dogen: Mystical Realist* (Hee-Jim Kim 1987) and translations such Cleary's *Rational Zen* (2014). Titles emphasizing realism and rationality are wonderfully modern but do not convey the non-rational world view of Feudal Japan where the icon of philosophical Zen is cured of serious illness by a parochial god (Williams 2003, p. 266). These contrasting versions of Dogen and Zen are illuminating. The uncovering of a pre-modern or at least less-than-modern Ch'an/Zen delineates historical accuracy from contemporary self-representation. That Zen was far more pre-modern than Shaku Soen and D.T. Suzuki would have us know only puts in vivid relief the modernist campaign to represent Zen as rational and scientific. Ch'an/Zen did not come modern from the T'ang dynasty; it was rendered modern by Soen, Suzuki, and others and eagerly embraced as a rational religion by new converts.

Natural Science and Psychology

Soen Roshi's talk at the World Parliament of Religions was entitled *The Law of Cause and Effect, As Taught by the Buddha*, aligned Zen with science and the Western Enlightenment. Roshi's presentation also put Zen in step with Sri Lanka's Protestant Buddhism as represented by Dhammapala. Both men emphasized the compatibility of their traditions with modern science, especially when compared with Christianity. However, with the waning of the adulation of science, especially among those attracted to Zen in the West, the affinity with science receded as a self-presentation theme for Zen. The contention that Zen is scientific was never heard by the researcher. Zen teachers, however, do retain openness to scientific analogy and metaphor more so than in the Forest orders.

Perhaps the most enthusiastic borrower from science is Roshi Nishijima, a contemporary Soto monk. Nishijima explains zazen, the core practice of his sect, as a switch between the sympathetic and parasympathetic nervous systems (1994, p. 55). A neurophysiological model of taking the seat of the Buddha is constructed. Certainly, this reductionism leaves much out, at times to the embarrassment of his students, but Nishijima's work is marvelously concrete antidote to any airy notion of spiritual practice. In his creative exposition of Dharma, Sasaki Roshi was not adverse to scientific metaphor and analogy. In addition to the already mentioned Buddha is the center of gravity equation, Sasaki, in a series of talks attended by the researcher at Holyoke College, employed a variety of science-based analogies to explain duality and non-duality. In neither case of the two teachers mentioned did they equate Buddhism with science. The more restricted use of science conveyed the notion that Zen was modern but was not being reduced to science.

We need to be careful about generalizing on modernism in Zen. The forms of modernization wax and wane, with earlier teachers and Asians perhaps quicker to draw parallels with science while late twentieth century teachers and Westerners being more circumspect. One can speculate that the science=Zen equation is less attractive to today's audience, and that the teachers are more sophisticated about the respective thought systems of science and religion and therefore refrain from facile analogies. The researcher had practice experience with at least two Western Zen teachers who had formal education in the sciences. One teacher was a Ph.D. in mathematics with a career in the aerospace industry. The other Zen teacher was a former professor of biology. Yet to the author's knowledge neither imported science into the Zen teaching.

There is a similar progression from zeal to more measured thought in regard to psychology. Early efforts such as Suzuki's *Zen Doctrine of No-Mind* (1972) and his collaboration with Erich Fromm, and DeMartino's entitled *Zen Buddhism and Psychoanalysis* (1960) have given way to works investigating the precise relationship between Zen and psychotherapy. From grand broad brush, theoretical interest has turned to the applied operation of techniques to healing the psyche. The work of Barry Magid (2000), a practicing psychoanalyst and Zen teacher in the Maezumi/Joko Beck lineage, is indicative of the newer generation of Zen/psychology interface. Inquiries used in the analytic context find their way into the Zen interview, but the exchanges are specific and do not reduce one discipline to the other through general theory.

However, the psychologizing of Zen is not limited to academic or popular literature. Psychology's mix with Zen is tacit in the culture of western Zen centers. Psychological thinking and language is ubiquitous if at times unsophisticated. Many Zen students come to practice with intellectual knowledge and some with psychotherapeutic experience from both sides of the couch. It was common to hear of students at Zen centers being enrolled in programs of psychology-related fields. A dharma heir of Korean Zen Master Sueng Sahn for instance undertook counseling training after being certified as a Zen teacher. While the researcher was active in the New York Zen Community, several members were practicing psychotherapists and others pursuing master's-level degrees. In fact, the complaint that was heard was that the teacher was not sufficiently psychological. A number of students bemoaned that the teacher in the Zen Community of New York lacked the psychological background to adroitly manage difficult persons in the Sangha.

The highly psychologized environment of Western Zen is easily ignored as it replicates the larger culture, functioning as wallpaper that does not garner attention and yet is the omnipresent background for group consciousness. Psychologizing, however, was not merely a shared culture for Western Zen students but was an actively applied interpretation of Dharma. Again, using the Zen Community of New York as an example, when a newcomer objected to the presence of a Buddha statue on grounds of idolatry, the issue was psychologized. Roshi Glassman's retort was that Zen was about smashing the idols we worship in the mind. Point taken and yet the ground of discussion was undeniably shifted from the external to the internal realm. Likewise, rebirth was interpreted as a moment-to-moment psychological fact rather than an after-death reality. The redirection to the interior space of the actor points to two other hallmarks of modernism: de-ritualization and the privatization of spiritual experience.

Forest Movement teachings, despite the education level of its Western monks, are largely devoid of scientific explanations or analogies for the content of Buddhist thought. Nevertheless, the Forest Tradition presents itself as compatible with the scientific approach. Science is referenced not as an explanation of specific practices but as a general philosophical outlook. On several occasions, Buddhism was presented as a rational inquiry or experiment. Meditation was an investigation carried out in the laboratory of the body/mind. The Forest Movement seemed more interested in the scientific method which is in keeping with its emphasis on investigative insight.

De-Ritualizing

A corollary to de-mythologizing is de-ritualizing or the simplification and marginalization of ritual in the practice. For the mass of lay Buddhists in Asia, the practice is ritual performance. In nearly two decades of living and practicing in Asia, it was rare to encounter a Buddhist, especially in rural areas, who had acquired anything beyond the most rudimentary grasp of Buddhist doctrine. Buddhism was ritual performance that earned merit. On bringing American college students to local villages, their lack of proper decorum was repeatedly commented on, but villagers seem to have little interest in the dhamma teachings that had been offered. Buddhism consisted of a few primary doctrines. Karma: do good get good, do bad get bad and impermanence: we all die was the typical extent of known Buddhadhamma. On the other hand, the external behavior that gained merit was carefully taught to children and enacted by adults. When the researcher inquired about Buddhism with school children, one was taught how to bow.

The modern demands a shift away from ritual through both simplification and marginalization. In the West, the Thai Forest Tradition has tended to preserve its rituals and liturgy more than the Ch'an/Zen groups, but not without some modification. The Thai Forest order deems the discipline prescribed in the Vinaya as essential to its 2500 survival, and no doubt this accounts for its more faithful replication. Nevertheless, modern modifications have occurred. One of these inroads by the modern shared by Amaro Bhikkhu is the leading role of ordained women in the chanting. He describes Ajahn Chah's start on one of his stays in England at hearing female voices in the foreground of liturgy recitation.

Perhaps the most evident area of de-ritualization in Forest Monasteries transplanted to the West is the ordained/lay etiquette. The typical Thai posture of sitting with legs bent to the side when before a monk becomes a more laissez-faire enactment accommodating Western inflexibility as long as feet are not pointed at the monk. The prescribed three bows or head to the floor prostrations are reduced to a simple wai or hand pressed together bow from the waist. In general, the demeanor is less reverential and hierarchical. A forest monk reports Ajahn Chah''s bewilderment at this apparent lack of ritualized respect coupled with the obvious interest of his audience in his teaching.

The importance of de-ritualization in this regard is not to be underestimated. The Theravada monk is an object of merit making and the robe is held in sacred esteem. An incident early on in my stay in Thailand illustrates the point. Subsequent to my presentation at a meeting, a monk approached me for further discussion. We sat together chatting. Thai laypersons were shocked as he had solicited my opinions and we sat at equal heights. On the occasions when doing solo retreats I would simply sit across the table when discussing practice with the abbot of the forest retreat center. Upon the arrival of Thai guests he would gracefully rearrange our relative positions to conform to local expectations. Modernism with its democratic, egalitarian ethos had invaded the forest but only so far.

If the Forest Tradition has successfully retained its ritual, with some pruning, then Western Zen has engaged in more radical cutting and grafting. The liturgies in centers such as Indiana Soto Sangha have kept to the original as found in Japan with English translation. Other centers have taken a more syncretic approach drawing on other Buddhist traditions to fashion a universal liturgy. Bodhi Zendo, for instance, a Zen center in India founded by a Jesuit priest who trained in the Sambo Kyodan sect, draws its chanting not only from recognized Zen Mahayana sources but also from the Pali. Chanting in English has been a critical shift in Zen centers; one not always appreciated by elders. Roshi Kapleau (Prebish 1999, p. 18) reports his dispute with his teacher Yasutani Roshi regarding dropping Japanese for English in chanting the *Heart Sutra*. The latter's contention that Japanese has a quasi-magical power not present in English is a pre-modernist response to a modernizing process. The Dharma at the practical/ritual level is literally being not only translated but transformed. Reportedly, this dispute led to the break between master and student.

Second only to chanting practice, bowing practice is the area of most recognizable modernization. In Zen, as with the Forest Tradition, the ritual aspects of monk/lay and teacher/disciple etiquette have been simplified. In place of the three bows performed across Asia, a single half bow often suffices. At no Western center did the researcher have to perform the three prostrations typical in the East. One entered the interview room gave a sign of respect, which in the Mahayana tradition was returned, and took one's seat. The abbreviation of bowing by no means meant it was now inconsequential. Shunryu Suzuki Roshi speaks eloquently of bowing (1970). The volume *On Zen Practice* (1976) coming out of the Zen Center of Los Angeles pays considerable attention to the practice.

Sheng Yen considered bowing to be a foundational practice and full prostration periods often concluded the day on long retreats. Yet private interviews commenced and ended with a single bow, conveying a modern attitude of dispensing with formality and getting down to business. Gouyuan, the monk in charge of practice at Dharma Drum Mountain, pointed to Sheng Yen's simplification of prostration as a modernizing adaptation. Shen Yen's modifications are not limited to de-ritualization of bowing practice. Although technically not a de-ritualization, he added elements that appeal to modern needs. A qigong practice of moving meditation is taught and practiced twice daily at retreats. Sifu has been heard to comment on several occasions that today's modern practitioner is "scarred" with stress, and that physical healing is a necessary corollary of Ch'an practice. This is an aspect of his Humanistic Buddhism and engagement with the ills of the modern world.

In general, the Thai Forest Monasteries outside Thailand, as opposed to vipassana centers, have sought to and succeeded in exporting their ritual tradition with fewer modifications. Vipassana centers, East and West, have tended to de-contextualize practice to a far greater degree than Forest centers (Schedneck 2012). In the West, Forest monasteries are unabashedly Theravada while insight meditation centers have an ecumenical, sometimes almost secular, atmosphere. The Forest Tradition retains its Theravada heritage and with it a more cohesive tradition than the Mahayana of Zen whose traditions are more national and less catholic. For instance, all Theravada centers chant in Pali while for the Mahayanists the Sanskrit is lost and replaced by the national languages. Change opens the door to further change as the shift to chanting in Western languages demonstrates. Finally, the Forest centers in the West are legally tied to their home monastery more so than Zen centers. While the Indiana Zen center and Ch'an Dharma Drum centers are linked with their Asian parent organization, many Zen centers are functionally and legally autonomous. Obviously, independence can translate into innovation.

Romanticism and the Non-Rational

Modernism contradicts itself and its dynamic is in part about reconciling those tensions. The triumph of reason that can be seen in de-mythologizing and de-ritualizing is brought up short with the romantic reaction to the mechanical world of getting and spending, to use Wordsworth's phrasing. Theravada and Mahayana modernizers celebrated Buddhism compatibility with science but they also celebrate non-rationality and the appeal to feeling. Myth may have been out but the notion of a non-rational solution to life to be known through something other than linear logic remains and was often trumpeted. Romantic motifs undoubtedly appeal to the modern consciousness, but we should be cautious in concluding that these themes only derive from contemporary input. Images of nature, of the authentic man beyond the corruption of urban life, and of unconventional behavior have been selectively transmitted by Buddhist modernizers. The Zen of D.T. Suzuki has little to say about the routinized monastic life and much to say about the antics of T'ang dynasty rebels and mountain poets. Modern conveyors of Buddhism have amplified the romantic but Buddhism has its own romantic roots.

Thanisarro Bhikkhu sees the romantic elements in popular versions of Buddhism as deriving from Western romanticism. His understanding echoes Taylor's observations about the role of the Romantic Movement and the origins of the modern self (Thanissaro 2012). The well spring of Buddhist romanticism, however, may not be the Rhine but the Ganges and the Yangtze. The transformation of codependent origination from a frightening truth that spurs detachment to a vision of unity and holism begins in the Mahayana sutras not with Heine or Wordsworth. The encounter with Chinese thought and Daoism made nature and the man of the way central to Ch'an sensibilities as evidenced in its poetry and stories. Reaching back even further to the Buddha himself, we find romantic musings as he extols the forest and the life of solitude. The Awakened One tells us to "take delight in the forest" (*Nalaka Sutta*). In the Rhinoceros Horn Sutta, the Buddha expounds on the virtues of a life of solitude. In the Dhammadpada one is urged to "devote (oneself) to solitude." In the *Arana Sutta*, we hear of "Those living in the forest, Peaceful and calm" (Ireland 2006). Like modernism, Buddhism is multi-themed. There are the analytical lists but there is also the call of nature and aloneness where the Dhamma shines. These sentiments pre-date Rousseau and Thoreau.

The Thai Forest tradition and Ch'an/Zen bring to the yearnings of the present-day seekers their own romanticism of the mountains and forest. The often heard motto in Thailand that the Buddha was born in the forest was enlightened in the forest and died in the forest conveys more than an environmental coincidence. To be in the forest is to be in the land of the Buddha. The writings of Tiyavanich (1997, 2003) while presenting valuable historical information also invoke the exotic setting of dhamma devoted monks in the forest. The solitary monk confronting both inner and fears and outer dangers is an appeal not to the logic of Dhamma but to adventure and drama in the archetypical romantic landscape: the untamed wilderness. The forest teachers not only lived and survived the natural but learned from nature in a manner consonant with the sensibilities of European Romantics and New England Transcendentalists. Nature is a source of similes and analogies for Forest Teachers. Ajahn Pasanno has offered public talks and retreats on Ajahn Chah's reflections on nature.

Ajahn Lee sets aside the scriptures to declare nature as the great teacher, indeed as Dhamma itself. In the view of the Western counter-culture, the Thai Forest Movement is a back to the nature movement.

The romantic and non-rational aspects of Zen have been widely noted. The Ch'an tradition gifts the world its mountain dwelling poets and eccentrics who embody the revolt of the natural against the artificial. Indeed, the categories of sacred/profane, awake/deluded, seem to be supplanted by natural/artificial. Long before reaching the shore of San Francisco Bay, these qualities were parts of Zen folklore. Once on land they found fertile ground in the Beat revolt and the 1960s counter-culture. The protagonist of Kerouac's *Dharma Bums* (1986) works a summer job as a firewatcher, an employment perfect for fulfilling the Buddha's admonition to seek solitude in the woods. Gary Synder proclaims a rucksack revolution, an American style thudong. From a historical view, the analogies are imperfect but in the presentations of the Buddhist modernizers and the imaginings of its present-day enthusiasts the match is compelling.

In contemporary Zen, the affinity for nature is not limited to the aesthetic sensibilities poignantly present in Chinese poetry but extends to everyday lifestyle. Satellite institutions of Zen Center such as Green Gulch Farm, Zen Bakery, and Greens Restaurant produce, prepare, and serve organic, natural foods, a fare far from the gruel encountered at Chinese temples. The cult of the natural extends to personal choices in clothing and home decoration and even to the choice of incense. Ring of the Bone Zendo, a northern California center, has reportedly ceased burning incense because of its chemical content and possible carcinogenic effects. While neither Chinese nor Japanese Zen has, strictly speaking, a thudong tradition; hermits and solitary practice have long been part of their heritage and constitute not merely a mode of practice but have a romantic draw. The documentary *Amongst the White Clouds* (Burger 2007) mentions the hardships of mountain living but shows a world of silence, harmony, simplicity, and peace far from the pollution of Beijing. Several showings organized by the author attracted overflow audiences clearly entranced with this nether world in the clouds. Even Ch'an's independent documenters cannot refrain from romantic advertising.

Nature in the Romantic view, however, is ultimately an externalization of the inner nature. Outer nature inspires but inner nature is the ultimate source of non-rational truth. As much as Buddhists may proclaim their logically coherent doctrine, Nibbana or Awakening is admittedly beyond the calculations of mathematics or the laboratory test tube. Here in our trans-rational nature the resolution of the human quandary is sought. This is the romantic's faith and yearning. The romantic impulse, however, for most practitioners does not lead to the mountain hermitage but to the inner cloister. This privatization of spiritual life has been identified as a modernist theme by several scholars of Buddhism and modernism (McMahan 2008, p. 7). In place of ritual space and community, modern religious sensibility relocates to the interior of the self. The emphasis by both these Buddhist schools on the trans-rational solution, Awakening, aligns with the modernist privatization of religion. Awakening is the quintessential subjective moment that reveals the truth of life and the universe, and is the great Release from the worldly. Ajahn Thate tells us the *Only the World Ends* (2007). The Thai Forest Movement and Zen offer up a mix that matches the romanticism of modernism without a retreat into pre-modern forms.

In sum, both movements contain romantic elements that are prominent in their self-representation: forest, mountains, nature, solitude, and the personal quality of transcendence. Each appeals to the consciousness seeking escape from the mechanical, materialistic world of consumerist collectivism. Zen's flamboyant enlightenments and teaching antics may be better known, but the romantic resonance of the Forest Movement's call to live the life of the Buddha should not be underestimated. Again it should be noted that many of these elements pre-dated the encounter with the modern. Rather than a simple adoption of modern forms or misinterpretation by contemporary adherents, romanticism and transcendentalism were already present. We see a process of the selection and emphasis of these elements by the movements' transmitters that establish a creative rapport with the modern.

Engaged Buddhism: Ch'an/Zen

Engaged Buddhism, a term coined by Vietnamese Zen Master Thich Nhat Hanh, is a significant current in Buddhist modernism. Along with revisions in practices, and intellectual sensibility, the meeting with the modern has also entailed an entrance into the outer predominantly secular world. This involvement in social action has generally been subsumed under the rubric of engaged Buddhism. The Engaged Buddhist Movement seeks to apply both Buddhist teachings and practices to social strife and suffering. Buddhist luminaries such as the Dali Lama, Thich Nhat Hahn, and Buddhadasa Bhikkhu are its recognized leaders and formulators. Ch'an/

Zen has been thrust into the arena of social service and activism by among others Bernie Glassman, Robert Aitken, and Sifu Sheng Yen. The forest monk, Ajahn Jayasaro, has taken the forest into the field of education. If the romantic has turned us away from a world too much with us, then engaged Buddhism has reversed direction bringing the Dhamma back into the conditioned world. To be modern is to be in the world. Again modernism confronts us with its self-contradiction, and Buddhist modernism shares this apparent paradox of romantic flight and civil engagement. A Buddhism that advertises the otherworldly solution, enlightenment, finds itself responding to a modernist sensibility that calls for involvement in a socio-political world.

As with de-mythologizing, de-ritualizing, and romanticism, engaged Buddhism is a modernist trend found throughout global Buddhism. Engagement is certainly not peculiar to the forest monks or Ch'an/Zen but these cases do offer unique hybrids. They draw upon different sources of inspiration and have to integrate a new worldliness with their otherworldly traditions. Each of these movements has had to travel a distance from the peripatetic thudong monk to the education reformer and innovator, from the hermits of Cold Mountain to a Yonkers N.Y. construction company. Zen has an advantage but not exclusivity over its forest cousins in embracing the modern and its emphasis on this world. One can point to the Mahayana as a more world-friendly orientation. Laypersons are present in its annals and daily chanting in Zen temples features the heroic if not daunting opening lines of the Bodhisattva vow: "Sentient beings are numberless, I vow to save them." Zen has also been recognized as the Buddhism of action. As Sasaki Roshi would remind his students experiencing the Buddha world was not enough, we had to "m-a-n-i-f-e-s-t." Ch'an/Zen has a more self-reflective, theoretically elaborated position on the place of worldly projects in spiritual life. In an early essay entitled "The Ideal Society," Shen Yen comes to the conclusion that while the goal of Buddhism is world transcendence, the path/practice of Buddhism is world engagement. Such formulations within the Chinese Buddhist Humanism Movement draw on the Mahayana Bodhisattva Doctrine, giving it a modernist cast. While Buddhadasa has a conceptualization of Dhamma that promotes social activism, there is no such doctrinal development in the Forest Movement, which remains more persistently focused on the inner, individual pursuit of Nibbana.

Nevertheless, the author's experience with Thai monks found regular participation in village work projects, youth groups, and social services.

The image of Theravada monks as withered recluses is unsupported by their activity in everyday life. Also, the Theravada concept of *phuttaphum* is a rough equivalent to the Bodhisattva way and an alternative to the arahant ideal. However, those who followed this road among the forest teachers were not social activist but dhamma propagators. On the other hand, the Mahayana Bodhisattva ideal affords the Ch'an/Zen school theoretical prop, if not imperative, for its forays into the social world. However, along with doctrinal support the more powerful factors may be the economic development of Japan, the penetration of missionaries and colonists into China, and the more extensive exposure of their intellectual circles to the West that ultimately account for today's Ch'an/Zen's more robust interest in social action.

In both the Chinese (Ch'an) and Japanese (Zen), there was opening to the social work of Christian churches. Taxiu returned from Europe having seen Christianity in action, and social service became a part of his Buddhism for Human Life not just death rites. Under the rubric of Humanistic Buddhism, Dharma Drum Mountain, Sheng Yen's organization has launched numerous social projects, including responses to geographically distant natural disasters. The latest example is relief efforts for the Nepalese earthquake of 2015. Sheng Yen's larger vision is of building a Pure Land. Transplanting the Western Paradise to planet earth would be a complete relocation of the telos of Sino-Buddhism. Sheng Yen was not alone but one of four recognized modernizers grouped under the rubric Humanistic Buddhism and known as the Four Heavenly Kings of Taiwanese Buddhism. Collectively, Sheng Yen and his three dharma brothers, Hsing Yun, Wei Chueh, and Cheng Yen, constitute a major force for Ch'an based engaged Buddhism.

While taking us beyond our regional scope but not to be ignored is Thich Nhat Hanh's Order of Interbeing. Thich Nhat Hahn, a Vietnamese Zen teacher, is a seminal figure in both the theory and practice of engaged Buddhism. From his center in France, Plum Village, his organization promotes a variety social efforts primarily revolving around peace and the healing of the wounds of violence. Efforts range from the practice of conflict resolution to the treatment of torture victims from around the world. One of Thich Nhat Hanh's disciples, ordained nun sister Chan Khong, has established an identity in her own right as an activist, notably in being a primary mover at an international anti-slavery conference. Beyond activism, Thich Nhat Hanh's Order of Interbeing has a revised set of precepts that move past prohibition to the prescription of socially responsible living. At the conclusion of a 10-day intensive retreat at which the researcher was a participant, Sheng Yen invited the monks and nuns of Thich Nhat Hanh's order from a nearby monastery to join in the closing ceremonies, underscoring the affinity of Taiwanese-engaged Ch'an Buddhism and Vietnamese-engaged Thein Buddhism. Ethnic differences and at times open hostility were eclipsed by a shared Buddhist modernism.

As with Sino-Buddhist Ch'an, Japanese Zen finds its activist roots in the early twentieth century prior to its export to North America. The New Buddhism of Shaku Soen and his colleagues was stimulated by the example of Christian charity and activism and soon began to give a new orientation to Japanese Buddhism, especially Zen. With Zen's arrival in the USA, a land of unrelenting causes, Zen activism reached new levels of intensity. Perhaps the most energetic expression of Zen action emerged out of the New York Zen Community under the dynamic leadership of Roshi Glassman, an American teacher and the primary dharma heir of Maezumi Roshi. The author's participation in the formative days of this community allowed a firsthand view of the transformation of a traditional Zen center into a socially concerned business. Under Maezumi's urging, a property was purchased, fitting his notion of Zen. Stately buildings sitting majestically in the hills of Riverdale N.Y. perched above the Hudson River were acquired from Columbia University. The grandeur of American Zen was short lived. Within a few years the community had moved to Yonkers N.Y., a working class suburb, where its Zen construction company built low-cost housing. Greystone Bakery catered to N.Y.'s affluent and provided funds for social projects. Even Zen practice was transformed under the social consciousness of engaged Buddhism. Retreats were no longer cloistered but were held on the streets of the Bowery, where participants were challenged to survive on a dollar a day or at Auschwitz, reflecting on the horrors of our world. The distance from the hermits of Zhongnan Mountains could not be further.

Engaged Buddhism: Thai Forest Movement

The journey of the Thai Forest Movement into the twenty-first century and an increasingly globalized world is convoluted and multi-directional. A tendency within the Thammayut nikaya promulgated by Bangkok elites with their modernist pretentions, the Forest Tradition, nevertheless, found its true inspiration in Buddhist orthopraxy and forest ascetics. With the end of WWII the elites finally recognized the opportunity that forest masters offered for the establishment of major temples and for the center's extension into the foliage of the margin. This begins a paradoxical movement of participation in the nation state by the forest monks with simultaneous localization of the tradition through the worship of relics and the production of amulets. A third phase commencing in the late twentieth century is stimulated by Western monks and students who combine liberal, global attitudes with a dedication to the soteriological project. Forest teachers such as Maha Bua, Ajahn Thate, Ajahn Toon, and Ajahn Chah travel to the West and establish outposts. The contradictions in this process culminate in the culture war over female ordination.

Whereas Asian figures were primary initiators of Ch'an/Zen modernism, there are no such corollary figures in the Forest Movement proper. Certainly Mongkut, his son Chulalongkorn, and his half-brother Wachirayan, infused modernist elements into the Thai Sangha. As we have seen, their modernism was not that of the forest dwellers. "Modern" Thais became interested in the forest monks. Ouay Ketsing gathered their stories and promoted their meditation, but performed no modernist re-orientation (Taylor 1993, p. 258). Yet within a few generations, the recluses of the forest were to be succeeded by educators with more in common with Montessori than Mt. Meru. The inspiration of this revolution is more directly Western. The modernism of Thai Forest monks owes more to its Western monks. Thai teachers, especially Ajahn Chah, were receptive to foreign influences in everything from Mahayana literature to establishing an international center, Wat Nanachat. These openings in a tradition rooted in the least-developed region of Thailand are remarkable, but it must be acknowledged that they were stimulated by Westerns whose willingness to live the life of forest dwellers gained the Thai ajahns' respect and interest in a larger world. East Asian Buddhism, on the other hand, began and, in some cases such as Taiwanese Ch'an, brought to a high-level Buddhist modernism with inspiration from Asian sources.

Perhaps the most prominent Forest project has been in the education field. A decade ago, foreign monks from Wat Nanachat began volunteering time in local schools in the Ubon Rachatani area. The Thai system tracing itself to the reign of Chulalongkorn the Great is more quantity than quality. Near universal education is crippled by rote learning, ill-trained instructors, and formal and distant teacher–student/teacher relations. In an effort to improve education and move it in a more modern direction, Western monks have offered their time and advice. This could be seen as a new phase of the forest monks civilizing of the Thai margin. The early phase consisting of the campaign against ghost worship is now supplanted by the effort to modernize education. What Ajahn Sao and Mun would have made of thudong monks evolving into classroom teachers one can only imagine.

The most ambitious and to date successful forest monk-assisted project is the establishment of a Buddhist school for children in the Chiang Mai area. Panyaden is a multi-grade bilingual school staffed by Thai and native English-language teachers. Unlike the local schools in Isan, Panyaden's students are drawn from the foreign community and the Thai professional and business classes. Teachers both Thai and foreign attend meditation retreats with forest lineage monks. "Buddhist values" are taught and seen as guiding school operation and student development. A founder and advisor is Ajahn Jayasaro, an American born, prominent figure in the Forest Movement from the Ajahn Chah sub-lineage.

The entrance of religion into education is hardly surprising, but the evolution of a meditation school that has practiced and idealized the ascetic monk wandering in the wilderness in pursuit of the Unconditioned is an arresting development. Despite the otherworldly image the monk retains an identity as an educator in Thailand, and this is especially true for foreign monks and those of the more strictly trained Thammayut. For example, the head resident monk at a Thammayut retreat center associated with the Forest Tradition in a remote Northern village was called upon to teach in the local government school. The subjects were overtly Buddhist, but, as he confided to the author, for a monk dedicated to meditation and solitude this was a challenge. Of note is that the non-forest monks in the area were not called upon. The revivalism of the forest monk is not only orthopraxy but also, at least in the eyes of local educators, orthodoxy. The forest monk is properly schooled in Dhamma. In regard to Panyaden School, the framing of the education as "Buddhist" seems to be an effort to bridge the forest ascetic and the modern educator. Whether by invitation or invasion, the modern world has come to the forest, but the forest has also emerged to meet its alien guest.

The term engaged Buddhism conjures up social service work such as disaster relief efforts or health care projects. Yet, Buddhism has an extensive and not always savory history of political engagement, as well. From the court intrigues of medieval Tibet and Japan to the anti-colonial and anti-militarist activity of Sri Lanka and Myanmar, the Sangha has displayed a persistent interest in the political. Despite its ascetic praxis and otherworldly telos, the Thai Forest Tradition, or at least one of its luminaries, has been a vociferous voice in contemporary Thai politics. Maha Bua in the wake of the collapse of the baht launched his Help Thai Nation Project which raised 12 tons in gold bullion to try and save the imploded currency. The government of Chuan Leepak was less than enthusiastic, so the arahant supported his rival, Thaksin, in the next election. In perhaps an unintended teaching on the principle of inconstancy; Maha Bua unequivocally repudiated Thaksin in 2005, attacking the prime minister personally and publically. Thaksin responded by suing the host publication for libel. Whatever the financial merits and ethical accuracy of the Maha Bua's positions, the forest master had come out from the trees and was now a player in national politics. If his fellow monks had not shaved their eyebrows, they might have raised them at a monk who is forbidden to handle money and barred from voting involving himself in such worldly matters.

Maha Bua's Thaksin and gold bullion affairs raise several points. Maha Bua's entrance into contentious politics is unique in the forest brotherhood. While the researcher heard similar sentiments among his Thai monks, they chose a low profile on the matter. On English-language websites and publications, the episode goes unmentioned. Next in terms of modernity, Maha Bua position suggests a romantic, reactionary response. International currency markets and upstart tycoons are the problems and Buddhist orthopraxy the solution. The Thai Forest Tradition's revivalism despite its modernist aspects also has anti-modernist reflexes rooted in its staunch ethicism and in some cases an affection for socially conservative forms. Yet we must be cautious and remember that Maha Bua is not the designated spokesperson for the Forest Tradition, even as his outspoken opinions may find sympathy among his brothers. There are other currents in the forest stream.

The monks centered around Ajahn Chah, especially Westerners at Wat Nanachat, represent a more liberal attitude. Chah himself as already indicated found himself intrigued, delighted, and educated by the culture peculiarities of Westerners. However, aside from the interest in education culminating in the Panyaden School, engaged Buddhist activities have been limited among Chah's monks as well as other forest sub-lineages. Monks at one of the smaller forest wats did coordinate with locals and petitioning government authorities to stop logging. Notably the action was defensive as the felling of trees was destroying the neighborhood immediate to the wat. The response was self-protective rather than an ideologically driven crusade. Conversation with forest monks generally revealed a favorable disposition toward ecology and peace issues, two of the most active areas of engaged Buddhism, but there was little in the way of social activism. What superseded social activism on the forest monks' agenda was the commitment to personal liberation. This was true not only for Thai monks but also for Western monks living in the West. Western monks held progressive attitudes on social issues that were held in check by a commitment to an otherworldly soteriological project. The suspicion with which forest monks were treated first as heretics by the Bangkok center, then as Bangkokian agents by the locals, and finally as Communist sympathizers during the 1970s insurrection, derived not from their engagement but from their dis-engagement. The forest monk, with some exceptions, is an advocate of dis-engaged Buddhism.

Facing the Feminine

The most contentious issue for Buddhists in general and the Thai Forest Movement in particular is the role of women in Buddhism and female ordination. The modern is a cluster of values, a mind-set. Picking up one strand brings along others. Reason, romantic individualism, and psychologism are connected to egalitarianism not by strict logic but by participation in a shared *Weltanschuuang*. If the Forest Movement had hit all the right modernist notes for its foreign adherents and public, then with the issue of the place of women they were out of tune. The full ordination of women as bhikkhunis, the almost complete equivalence of monks, is a 2500-year-old question that has recently taken center stage within the Thai Forest Movement. The status of women is under modernist and feminist scrutiny in all branches of Buddhism, but the current descendants of the forest dwellers more than any other grouping find themselves facing present-day feminist challenge. Certainly no issue has exposed the limits of forest modernism and its traditionalism and enclosure by the Thai Sangha and state.

The background to the issue is complex. Briefly, it can be traced to the status of the bhikkhuni order as established by the Buddha and the technical conditions for its continuance. The Buddha's prescription of additional precepts for the nuns and his subordination of the nuns to the male order (no female ordained may criticize a monk or walk ahead of them) is interpreted by conservatives as indicative of the Buddha's reluctance to admit women to the Sangha. In Theravada Buddhism, the lineage of the bhikkhunis died out, and according to one strict interpretation of the Vinaya, the conditions of its reinstatement are no longer and will never again be present.

The present situation of female ordination varies across the denominations and nations of Buddhism. According to Sakyadhitya Daughters of the Buddha (http://www.sakyadhita.org), full and undisputed ordination can be found only in China, Korea, and Vietnam, three nations in which Ch'an is predominant. In short, Ch'an/Zen enters the twenty-first century without the legal/historical impediments to female ordination that the Forest Movement and other Theravada carry. In the Mahayana Sister Chan Khong (Vietnamese) and Soen Master Deahaeng (Korean) are examples of attained and fully ordained women. Women have also been sanctified in the Forest Tradition. State Buddhism devalued the lay practitioner both male and female, but forest teachers often praised women followers. Mae Chee Kaew, whose arahant status is memorialized by her stupa, and Kee Nanayon, who while not technically in the forest lineage was, according to a close associate of Maha Bua, acknowledged by the Master has having attained total liberation. However, neither of these women was fully ordained with Kee Nanayon remaining a laywoman. Therefore, it would be wrong to claim that the Forest Movement gives no place to women, but when it comes to bhikkhuni ordination, the line is drawn across the forest path.

The issue was being managed by forest monks and abbots with some difficulty when with the actions of Ajahn Brahm it ruptured into open crisis. On 22 October 2009, Ajahn Brahm ordained four women as bhikkhunis at the Bodhiyana Wat of the Western Australia. Up to this point, the most radical challenge to an all-male Sangha was posed by Dhammananda, a former academic who took full ordination outside Thailand, returning to establish an all-women temple and to be frostily ignored by the Thai Sangha. Ajahn Brahm's ordinations, however, were within the Thai Sangha and thereby threw down a gauntlet was that could not be ignored. The Ajahn Chah lineage and Wat Pah Pong, with which Ajahn Brahm was affiliated, had to respond. A clash was inevitable between monk privilege, Sangha rulings, cultural belief, and Thai law on one side and modern sensibilities and values along with Vinaya based arguments on the other.

Prior to the performance of full ordination in Australia efforts had been made to address the rising interest in ordination, especially among Western women. At Chithurst Monastery in England, Ajahn Sumedho, Ajahn Chah's foremost Western disciple, established, with his teacher's approval, a Buddhist monastic female order entitled *siladhara*. Formal permission from the Thai Sangha was received to give a ten-precept *pabbajjia* that conferred official recognition as renunciates in Chah's lineage, a status that mae chees and thilashins lacked. The women were trained in the patimoka-based precepts, bringing them closer in substance to male bhikkhus but not in status and privilege. The new order remained small, perhaps growing to 14 persons ensconced in England, but the worldwide pressures for full ordination continued. Counter-forces also mobilized to contain the ordination movement resulting in Sumedho's The Five Point Declaration which explicitly defined the siladharas as inferior to male monks. The effort to find the "ground between" failed as the declaration was repudiated as sexist and unflatteringly compared to the American Constitution's definition of the slave as three fifths of a person.

If Brahm's elevation of women renunciates to bhikkhuni was cause for celebration, it was short lived. Brahm was summoned to a meeting at Wat Pah Pong to account for his actions. After much rancor it was demanded that he disown his actions. He refused and was expelled from the world of Thai Buddhism. The limits of modernism had been reached. Thai Buddhism was to discover its own limits when Australian courts dismissed their efforts to retain control over the Australian Sangha and its properties.

Globally, no issue has ignited more dispute and impassioned exchanges. Websites and blogs such as New Buddhist, Buddha Space, Existential Buddhist, Wisdom Quarterly, Women in Buddhism, Dharma Wheel, and numerous others all featured conversations on the topic, most applauding the ordinations. Major Buddhist magazines, *Shambala Sun* and *Tricycle* in the West, gave space to Ajahn Brahm, and even the largest English-language newspaper in Thailand, *the Bangkok Post*, interviewed the now renegade monk. The dispute was not limited to the popular media. Thanissaro Bhikkhu, a luminary in the forest lineage, produced a scriptural refutation of the ordination which provoked counter-arguments from eminent scholar monks such as Bhikkhu Bodhi (2011) and Analayo (2011). The Thai Theravada world and especially the Thai Forest Tradition were now in dispute, at least in the eyes of their foreign audience.

Rather than seeking to adjudicate the textual and legal differences, the larger cultural dynamics at play will be delineated. The traditional culture and legal framework within which the Forest Movement operates was now exposed to its Western enthusiasts. Political forces within the Sangha and government had enacted laws specifically banning female ordination and attached significant penalties for violators. Thai temples, especially in the North, ban women from certain areas as menstruating polluters. What had been kept out of the awareness of Western students was now on full display.

Other than technical arguments based on scripture, there was little contingent communication as each side reiterated its non-negotiable views: one traditional, one modern. The common attitude encountered from Thai monks within the forest lineages was that women did not need ordination in order to practice. Westerners appealed to equality and compassion. The dispute took on a personal tone that reflected strongly held cultural values. Exchanges became ad hominem attacks. Brahm was an impulsive, loose cannon. Ajahn Kevala, an ordination opponent at Wat Pah Phong, was a scheming misogynist. Headlines declared that Brahm had been excommunicated conjuring up images of Inquisitorial persecution. Monks such as Amaro privately acknowledged that his was their most trying moment. Western admirers were compelled to re-vision the Forest Movement if not all of Buddhism. The culture wars had come to the forest.

The prominence of women in the Western Zen (and vipassana, insight meditation movement) contrasts with the Forest Tradition. The availability of full ordination comes to immediate notice, but the place of women within the Ch'an/Zen world is a larger phenomenon than ordination. On the one hand, women in the West remain sensitive to covert sexism. A survey revealed that most Buddhist women believed more could be done for gender equality in American Buddhism, but paradoxically most found no evidence of sexism within their own Sangha (Coleman 2002, p. 139). In contrast the Asian complaint has been of a shortage of practice opportunities and material resources even for fully ordained women.

The most conspicuous difference between women in Western Zen and the Western Forest Tradition is not simply ordination but general status. Zen's earliest days in America featured lay women who provide key organizational and material support, who then went to Japan to practice in temples that were opened to foreign women, and who were followed by women who ordained, received permission to teach, and acceded to the head positions in Zen centers and temples. The idea of an abbess presiding over the training of monks, such as Jiyu-Kennet of Mt Shasta Abbey, is unthinkable in the Kammatthana world. A quick review of teachers listed on the Soto Zen Buddhist Association website (http://szba.org/) shows just under half as female. If women have not yet attained their quota they are close. In contrast, one must look carefully for a female face on a forest monastery website. On the role of women, Zen and the Thai Forest Movement are contrasts. In short, in two areas critical to Buddhist modernism, social activism and the role of women, the Thai Forest Tradition and Zen, especially Western Zen, are markedly different. Both share elements of a modern social consciousness, but the forest monks have resisted its application to socio-political and gender domains.

Summary

Ch'an/Zen and the Forest Movement share a number of modernist forms: a rationalist doctrine that is open to science, a romanticism of nature and the non-rational experience, a de-mythologized and de-ritualized practice, and a new interest in engaging society. They also display differences with Western Ch'an and Zen distancing themselves from their Asian roots more than the Theravada Forest Tradition, and thereby opening up more innovative space in liturgy, ritual, and for the inclusion of women in positions of power. There are differences also in how they come to modernism. Ch'an/Zen found modernist notions among its Asian teachers who formed Asian-based Buddhist movements prior to their export to the West. The Forest Tradition was directly inspired by Western monks who came to Thailand. The issue of female ordination also corrects the analysis of the Forest Movement as a pragmatic-soteriological project by bluntly displaying the power of the state-run Sangha and historical tradition to impose hard limits. Acknowledging these distinctions and qualifications, both movements are, nevertheless, living experiments in a new, globalizing Buddhadharma.

References

Analayo, B. (2011). *On bhikkhuni ordination*. Retrieved March 3, 2015, from https://snfwrenms.wordpress.com/2011/11/23/bhikkhu-analayo-on-bhikkhuni-ordination/

Bechert, H. (1984). Buddhist revival East and West. In H. Bechert & R. Gombrich (Eds.), *The world of buddhism* (pp. 273–285). London: Thames and Hudson.

Bodhi, B. (2011). *Revival of the bhikkhuni ordination*. Retrieved March 4, 2015, from http://inwardpathpublisher.blogspot.com/2011/12/this-is-wonderful-slideshow-of.html

Burger, E. (Producer/Director). (2007). *Amongst the white clouds* [Video]. USA: Independent.

Cleary, T. (2014). *Rational zen*. Boston: Shambhala Press.

Coleman, J. (2002). *The new Buddhism: The western transformation of an ancient tradition*. Oxford, MA: Oxford University Press.

Fromm, E., DeMartino, R., & Suzuki, D. T. (1960). *Zen Buddhism and psychoanalysis*. New York: Harper and Row.

Gombrich, R., & Obeyesekere, G. (1988). *Buddhism transformed*. Princeton, NJ: Princeton University.
Hee-Jim Kim. (1987). *Dogen: Mystical realist*. Somerville: Wisdom.
Hughes Seager, R. (2006). *Encountering the Dharma: Daisaku Ikeda, Soka Gakkai, and the globalization of Buddhist humanism*. Berkeley, CA: University of California Press.
Ireland, J. (2006). *Aranna Sutta*. Retrieved March 4, 2016, from http://www.accesstoinsight.org/tipitaka/
Jackson, P. (1993). Re-interpreting the Triaphuum Phra Ruang. In Buddhist trends in Southeast Asia, T. Ling (Ed.). Singapore: Institute of SE Asian Studies.
Jayasaro, A. (2008). *Ajahn Chah's biography by Ajahn Jayasaro by the noble path*. Retrieved January 8, 2015, from https://www.youtube.com/watch#1-43
Kerouac, J. (1986). *Dharma bums*. New York: Viking Penguin.
Lopez, D. (2002). *Buddhism and science*. Chicago, IL: University of Chicago Press.
Magid, B. (2000). *Ordinary mind: Exploring the common ground of Zen and psychotherapy*. Somerville, MA: Wisdom.
McMahan, D. (2008). *The making of Buddhist modernism*. New York: Oxford University Press.
Nishijima, G. (1994). *To meet the real dragon*. Dogen Sangha Publications.
Pitman, D. A. (2001). *Toward a modern Chinese Buddhism: Taixu's reforms*. Honolulu: University of Hawaii Press.
Prebish, C. (1999). *Luminous passage*. London: University of California Press.
Sheng Yen. (2010). *The dharma drum lineage of Chan Buddhism*. Taiwan: Sheng yen Educational Foundation.
Suzuki, S. (1970). *Zen mind, beginner's mind*. New York: Weatherhill.
Taylor, C. (1989). *Sources of the self: The making of modern identity*. Cambridge: Cambridge University Press.
Taylor, J. L. (1993). *Forest monks and the nation-state*. Singapore: Institute of Southeast Asian Studies.
Thanissaro, B. (2012). *The roots of Buddhist romanticism*. Retrieved September 8, 2015, from http://www.accesstoinsight.org/lib/authors/thanissaro/rootsofbuddhistromanticism.html
Thate, A. (2007). *Only the world ends*. Bangkok: Pattanasuksa.
Tiyavanich, K. (1997). *Forest recollections*. Chiang Mai: Silkworm Books.
Tiyavanich, K. (2003). *The Buddha in the jungle*. Chiang Mai: Silkworm.
Williams, D. (2003). How Dosho's medicine saved Dogen. In B. Faure (Ed.), *Chan Buddhism in ritual context*. London: Routledge Curzon. Chapter 8.
ZCLA (1976). *On Zen practice*. Los Angeles, CA: ZCLA Publication.

CHAPTER 10

Conclusion

Introduction

Our working concern was to answer the initial question: How could two religious phenomena separated by geography, culture, history, and even denomination manifest similar forms? The question sets additional tasks. The assumption that there are affinities to account for must be demonstrated and the precise nature of the similarities elucidated. In the course of our study, we have sought to avoid comparisons that ignore marked differences. Indeed, the interplay of likeness and difference raises issues of what factors seems to promote differences and which support similarities. These are questions which will be addressed here in our final chapter. This discussion will also question the framework through which phenomena like Ch'an/Zen and the Thai Forest Movement are typically approached. In essence we will question the limits of established modes of Buddhist and religious studies and suggest an alternate view. Finally, we will speak to a set of concerns often exiled from formal studies: the needs and questions of practitioners. A study of the comparative characteristics of these schools is pertinent for those considering or engaging in practice in either. However, before addressing these themes it is appropriate to review the findings and explanatory models that have thus far emerged.

What came to light in our study is the central and shaping role of each movement's core orientation and its immediate expression in action and face-to-face relationships. Both Ch'an/Zen and the Thai Forest

A.R. Lopez, *Buddhist Revivalist Movements*,
DOI 10.1057/978-1-137-54086-7_10

Movement are animated by a shared impulse. The project has been termed pragmatic-soteriological. This hyphenated concept links together the ultimate telos, liberation, and its result-oriented methods. Given the commitment of early Buddhism to awakening, a soteriological project might not seem unique, but as both movements make clear, and is to some extent supported by the evidence, much of the Buddhist apparatus in China, Japan, and Thailand is engaged in other endeavors. A return, or in their understanding a continuation of Buddhism's inceptive interest, defines these schools as revivalist. Zen locates its continuity in a mind-to-mind transmission. The Thai Forest School rests its connection on the precision of its discipline, orthopraxy. Taken together, they can be defined as revivalist ventures driven by a pragmatic-soteriological undertaking.

If the term soteriological tells us the goal, then pragmatic tells us the way. It is pragmatic not only in its concern for results but also in its implied immediacy, in this very life, and its willingness to subordinate other values to this end. The pragmatic enterprise of the mountain and forest monks has two operating forms which as a cluster shape these schools. Self-cultivation and charismatic authority channel the soteriological enterprise. The individual actor is to be transformed under the guidance of the charismatic teacher. A simple imaginary-analytic exercise of substituting variables demonstrates the decisiveness of these forms. If liberation was attained via social works or sacrament requiring no charismatic teacher the entire temper and dynamics of the revivalist movements would be altered. Taken together, self-cultivation, charismatic authority, and working in the service of a pragmatic-soteriological venture constitute the factors that configure Ch'an/Zen and the Thai Forest Tradition. These inter-locking elements constitute not only its core social form but also the source of an inner logic that shapes the action, style, and tone of both movements. Given their similar commitment to charismatic teachers and self-cultivation in quest of liberation in this very life, an explanation of their independently derived similarities becomes possible.

Review

A review of our findings will allow for further development of a model of the factors that lay behind the similarities/differences of Ch'an/Zen and the Thai Forest Tradition. Each movement was identified as a revivalist reaction to the Buddhism of its place and time. Their self-presentation shares a claim of purity and authenticity reaching back to the Buddha and separating them, at least in their own view, from a Buddhism lost in Imperial-funded scholasticism in the Ch'an case, and in the case of the forest monks from a degenerate Buddhism on one side and a bureaucratized Sangha on the other. While this version can be disputed historically, it is the genesis narrative of each movement and their self-representation. Along with their similar positioning with respect to the larger Buddhist community, the central place of a pragmatic-soteriological project enacted through self-cultivation and charismatic teachers was noted. These variables are returned to again and again as critical to understanding the content and forms of the dynamics of both schools.

Turning to the substance of Ch'an/Zen and the Thai Forest Movement, several features catch the eye as points of affinity. Perhaps the most prominent is the unconventional, face-to-face teaching style. What is discovered is that despite their adherence to basic precepts, the imperative of self-cultivation and direct knowing, and the charismatic authority granted their masters injects a non-normative tendency into the teaching enterprise of both movements. Specific types of teaching tactics were identified along with their functions, and examples drawn from both schools. Both groups practiced utilization, modeling, confrontation, paradoxical demands, self-contradiction, and compassionate torment to shock, confuse, and pressure the student into trans-rational responses and insights.

The soteriological project means walking the path through the practices of self-cultivation. For both, meditation is the royal road to awakening. Here the shared aspects of meditation as practiced in the two schools were identified as well as key differences. Concentrative exercises laid the foundation, but investigation, especially of the body along with walking meditation, was emphasized by the Forest Tradition, while Zen undertook a direct assault on the nature of mind through either koan practice or "looking into the essence" of mind. The key differences appeared not in techniques but in their overall organizational style which were characterized as individualistic in the Forest School, and collectivist in Ch'an/Zen. On the other hand, the "spirit" or motivational styles of their paths echo one another. The controlling metaphors are martial. The quest for awakening is a campaign waged against ignorance and the seeker is a warrior of the way.

Next, we turned to awakening, the *raison d'être* of the mountain and forest schools. This elusive yet cardinal phenomenon was analyzed at the levels of descriptive content, transformed structures, and social functions. While acknowledging the limited sample and the unsystematic descriptions, a phenomenological outline was tentatively sketched, suggesting that in both schools awakening was a non-image dependent (NID) state often consisting of an inversion of consciousness, non-duality, and radical dis-identification with the body. The role of the teacher in verifying the awakening and the charismatic bestowing capacity of the event were highlighted. Awakening fulfills the ultimate purpose of each movement and connects it with the root charismatic, the Buddha, and his seminal experience under the Bodhi Tree.

The pivotal role of the charismatic teacher dictates the importance of biographical and auto-biographical compositions which often assume a laudatory, hagiographical form. The accounts of the lives of the awakened perform multiple functions. The credentials of the charismatic are documented. Inspiration is awakened in the disciple. The structure of the story was shown to often follow the pattern of rites of passage, a near universal process of transformation found worldwide. The archetypal journey of the hero as formulated by Joseph Campbell is a hidden template in the life tales of both schools. The individual life story of the awakened is thereby connected with a cosmic story. For the Kammatthana the model life is the Buddha and the arahants, while for Ch'an/Zen it is a composite figure of Bodhidharma and the early masters. Within the hagiographies of the Forest Tradition, it was noted that those written by the Maha Bua lineage have a more traditional and laudatory tone than other forest teacher accounts which have a more naturalistic and modern mood.

While both The Forest School and Zen insist on practice and often dismiss or marginalize doctrine in their self-representations, they also promulgate doctrinal innovations which reframe and reorient their practice/path. The essential revolutionary formulation of the forest masters is the teaching on the pure mind. Pure mind teachings have been singled out as the most important instruction transmitted by Ajahn Mun. The realization of the pure mind is the consummation of the journey, liberation. Pure mind or its attainment is Nibbana. Additionally, the doctrine of an inherent, unconditional, permanent mind brings the Forest School into close alignment with Mahayana teachings on Buddha-nature and more specifically Ch'an/Zen ideas of original mind and mind essence. Textual support for pure mind or nibbanic consciousness was located in the Pali literature. This affinity between Zen and the Forest Tradition establishes core doctrine as an area of proximity between the two schools.

The teaching on pure mind is the most explicit and prominent conceptual innovation by the forest school, but it is not the only doctrinal shift made by the forest ajahns. Two areas were observed: first the understanding of right view and second the handling of three concepts: emptiness, nowness, and skillful means. These were shown to exert a shift on the relationship of path to goal and the status of right view. Right view as classically taught is the holding of correct truth propositions, under the influence of a pragmatic-soteriological project right view becomes non-attachment to views. The critical question is how free or freeing is the actor's position rather than how correct is the belief. Support for this radical view of views was found in the Pali Canon, although it is not the mainstream teachings. The teaching tactics of forest teachers indicate the ascendency of skillful means and intuitive wisdom (panna) over truth propositions. Once again, this brings them closer to their mountain cousins. Ch'an/Zen throughout its history asserts that right view is the one that is effective. At times, Ch'an/Zen espouses the no-view view. All views are impediments to freedom.

A more subtle shift, yet potentially one of great implication, is the reconfiguration of the relationship between path and goal. Classical Theravada formulation of the four stages of enlightenment culminating in arahantship constitutes a progressive linear path. This contrasts with Ch'an/Zen's insistence on pre-existing enlightenment (hongaku, (Jp.); ben jue, (Ch.); pon'gak, (Kor.)) and its sudden recognition. The linearity of path is not merely attenuated but is refuted. The Chah-Sumedho lineage approaches this Ch'an/Zen view first by Chah's de-emphasizing the importance of formal levels of attainment, and then by Sumedho's suggestion that the essence of awakened mind is present here and now. He encourages his disciples to not look elsewhere for liberation, but to turn to the already present qualities of mind. Ch'an/Zen concurs.

Lastly, the relationship with modernism was shown to be complex. Modernism itself contained themes in tension with one another such as a science and romanticism. Buddhism from Sri Lanka to Japan proactively embraced modernism and presented the Buddha's creed as utterly compatible with reason and science. Humanistic Buddhism and New Buddhism predated Ch'an/Zen's direct interchange with the West. This contrasts with the Thai Forest Tradition. Although distantly related to Mongkut's modernizing interests had to await the arrival of Western monks for a dialogue to be opened. Modernist tendencies of de-mythologizing, de-ritualizing, and romanticism characterize both schools, with Zen, at least in the West, taking these processes further than their forest brethren. A similar pattern holds with lay/monk and gender relations with a general easing of Asian formality and a softening but not ending of gender inequality. Again, Western Zen groups displaying a greater capacity for modernist reform than the Theravada-based forest order. The limits of the Thai Forest School's interface with the modern were examined with regard to its crisis over female ordination. The Thai Forest Tradition, despite modernist sensibilities, was less interested in social activism and had hard limits on female ordination and authority.

The key points of affinity can be summarized as follows:

1. Teaching tactics in techniques and function.
2. Practice committed to self-cultivated liberation and a militant spirit.
3. Awakening experience: the phenomenological, the structural, and the social functions.
4. Hagiography as archetypical quest, essential group literature, and charismatic social role.
5. Pure mind as pivotal doctrine and the object of awakening and the absolute.
6. View subordinated to skillful means, emphasis on emptiness, and conflation of goal/path.
7. Modernist tendencies of naturalism, romanticism, and de-mythologizing/de-ritualizing.

Broader shared dispositions are: pragmatic-soteriological project, self-cultivation, charismatic teacher/disciple relationship, and a revivalist stance.

Theravada, Mahayana: The Cost of Vehicles

There is a ditty that has circulated in Western Dharma realms that reads, "Greater vehicle, lesser vehicle, All vehicles will be towed at the owner's expense." What has been the cost of vehicles not only for their owners but for also those who dare to study them? The division between Theravada and Mahayana as areas of study begins with the earliest of Buddhist studies. Nothing could appear like a more inherent, common sense distinction. Several factors dictate that the separation of the two is logical and necessary. The first is history. Each denomination has its own narrative played out in a geographically separate region. They did share place and time in Sri Lanka and China but these are treated as brief interludes before each branch comes to dominate its territory. Given the insularity of the denominations and their coherent regionalism, the stance of separate areas of study is understandable. To tell the story of each requires no reference to the other.

Along with geo-history, the second factor is the power and exclusivity of language. The active tongues of the Mahayana are the East Asian languages: Chinese, Korean, Japanese, and Vietnamese. Behind them stands the root-mother language of the Mahayana, Sanskrit. The exclusive association in India with Sanskrit and Pali by the two primary branches of Buddhism assured their assignment to different department of study. Experts on Medieval Tibetan are unlikely to undertake works on the Theravada. While there are studies that cross the divide, such as translations of the Dhammapada that address versions in both languages, largely the separation of the Pali and Sanskrit studies continues.

The last century has bridged this divide. Interestingly, this has often been due to the interest of scholar monks. Rahula Walpola was one such monk who as a Sri Lankan modernist with an inclusivist orientation became interested in Nargajuna, the Mahayana's dialectician of emptiness, and the Far Eastern subject of this study, Ch'an/Zen (Walpola 1978b). As the twentieth century progressed, intellectual curiosity was followed by practice/community based exchanges. In the West, Theravada and Mahayana groups shared space and found themselves co-members of a minority religion. Yet, cross-denomination exchanges, of which this effort can be seen as an example, have had minimal impact on the organization of Buddhist studies. Academic departments and individual scholars define themselves as Theravada or Mahayana or Vajrayana. Alternately, research is carried out under the egis of national/ethnic studies for instance Korean or Tibetan. The consequences, while largely unintended, are significant. The advantages are a coherent historical-regional narrative, the use of primary sources, and in general keeping Buddhism contextualized. The "Theravada" and "Mahayana" are foundational frames that can profitably shape subsequent consideration. Yet, the yanas or vehicles are towed with expense. Only certain roads can be traversed.

This study has taken a comparative route often flirting with the hazards of a-historicism and de-contextualization. These risks, however, offer the rewards of identifying dynamics, the factors involved, and evident similarities that would otherwise be unattended to by studies controlled by the rubrics "Theravada/Mahayana" or "Thai/Chinese." In contrast, overly contextualized studies run the risk of being limited to description and an inability to analytically generalize. Ethnographies can be solid building blocks but a house is more than its bricks. In choosing to extract out of context these phenomena and defining them as social movements of revivalist type, the way is open to examine similarities (and differences). Questions and answers become possible that would be unavailable had each been examined only on its own turf.

An additional critical point is that taking formal denomination labels as controlling concepts runs a risk of precipitous deduction. Can we really deduce what Ch'an/Zen and the Thai Forest Movement are about by gross designations such as Mahayana and Theravada? If we approach our studies with a bias toward religious ideas and labels, pigeonholing phenomena in these categories follows. The filing system is flawed. It asserts that broad ideological categories based on philosophical positions and interpretations of disciplinary codes in India several millennia ago are adequate starting and finishing points of study. As we have seen that the Theravada and Mahayana designations can point to active, shaping influences. These general headings do denote differences in form and style of the mountain and forest monks. The "outer" forms such as official dogma, rules of ordination, lay/monk proscriptions were shown to be derived from the wider tradition. What has been termed their "inner dynamic" such as teacher/student gambits, the spirit of practice, re-interpretations of doctrine; all reflect the revivalist project and its method, a pragmatic-soteriological undertaking of self-cultivation.

In short, the inner is a tendency or pressure that pushes the outer for change while the outer puts a stop on inner generated tendencies. The two can collide as was shown in the issue of female ordination. An inner dynamic of the pursuit of liberation crashes into Thai Theravada Buddhism. However, the general category, Thai Theravada, is not assumed to be an effective explanatory frame any more than Chinese Mahayana could account for the dynamic of Ch'an. A comparative study requires and allows for categories that identify patterns that arise from the projects and the intentions of actors that reside behind denominational labels.

Recently, the author has a chance to view the crafts of Taiwanese ethnic groups. The similarity with hill tribe productions in Southeast Asia was obvious, yet these groups were from different nation-states and carried different official identity cards. If we asked about their citizenship only, we would miss the similarities of their constructed worlds. By asking Ch'an/Zen and the Thai Forest Tradition about more than their card carrying designations, we have elucidated shared tendencies that otherwise might be missed. The tribe as a working group has been emphasized over their formal citizenship in alternate vehicles. The Ch'an/Zen and the Kammatthana Forest Movement are of the revivalist and pragmatic-soteriological tribe and demonstrate tribal affinities.

For the Practitioner

The relationship between the forest and mountain is not just an academic affair. Indeed, the inspiration for this enterprise is a meditation retreat that drew Buddhist teachers, lay and ordained, from a variety of denominations. The discovery of unexpected overlap in teaching methods and doctrine energized an awareness and exploration of the interface of previously discrete "Buddhisms." At another retreat a participant shared her joy in coming back to the Mahayana from her more recently adopted vipassana practice. The Ch'an meditation as taught by Sheng Yen bridged the two traditions in her view. A multipart interview with this teacher published in *Tricycle* was conducted by a significant member in the Cambridge Insight Meditation Society, a Theravada vipassana based group. He had been practicing several years in the Ch'an style (Sheng Yen 2003, pp. 16–20).

There is a fluid movement between Buddhisms in the West. Justin O'Brien McDaniel, a scholar of Southeast Asian Buddhism, has coined the term a repertoire of practices to denote the tendency of Southeast Asians to employ a number of different modalities in their Buddhism over the course of a lifetime (2011, p. 9). The staid categories once again give way to a far more hybrid and at times indeterminate forms that convey the creativity of actors. Westerners, unconstrained by inherited insular traditions, are showing an interest in developing a repertoire. Attracted by the affinities between Ch'an/Zen and the Forest Tradition, a new Buddhist may find of value a discussion on the implications for practice. Choices entail discrimination between the differences of the two schools. Where they nearly replicate each other, there is little need in choosing one over the other. Mutual appreciation suffices. Where they differ yet also have common features, the alternatives may contribute and still coexist with a larger style.

Each has domains with alternative methods of practice, doctrine, and ordained/lay relationships. Here we view these differences from the perspective of the actor in practice. The uniqueness of each practice can now be seen against the background of the similarities that have preoccupied this inquiry. How does Buddhism live differently in the forest and the mountain? The answer can be deduced from the institutions of each movement, but we also have the observations of those who have directly and personally experienced working with each school. Attending one another's retreats, in some cases building a relationship with teachers from the other tradition, provide firsthand knowledge that non-existent a century ago.

The most meaningful contrast in practice is not at the level of technique but in the overall organization and philosophy. We have labeled Ch'an/Zen practice and temple life as "collectivist" and the Forest Tradition as "individualist." In moving from one to the other or sampling the two and then choosing a path, one is confronted with either a group structured practice or one that throws the seeker back on her/himself. The root similarity is that both traditions demand self-effort. Japanese Buddhism distinguishes two primary approaches to the role of human will power, self-power or *jiriki*, and other-power or *tariki*. The latter is typically associated with Pure Land/Amida sects, while the former is identified with Zen schools. In the self-power, you have to do it; no one will do it for you.

The Forest Tradition shares this understanding. Both Ch'an/Zen and the Forest Tradition stand on the classic pronouncement found in the Dhammapada.

> By self alone evil is done;
> By self one is defiled.
> By self evil is not done;
> By self one is purified.
> Purity and impurity are individual matters.
> No one can purify another.

(Roebuck 2010, p. 34)

Devotional forms of Buddhism which developed in the Mahayana and rely on the powers of Buddhas and Bodhisattvas, other power, are largely absent in Zen and the Theravada. There is suggestion of other-power in Dogen's Zen and in the Pure Land elements in Ch'an, but as transmitted to the West they are largely marginalized by reliance on rigorous sitting meditation. As do-it-your-self Buddhism, the practitioner finds a common critical orientation which locates at different points on an individualist/collectivist spectrum. The common commitment to self-responsibility manifests in a contrasting manner.

Contemporary enthusiasts are primarily driven by an interest in meditation and its possibilities for alleviating suffering, so considerations of the two movements as practice vehicles will commence in this domain. Northeast Asian Buddhism is organized around week plus long sittings that are regulated by bells and wooden instruments, collective eating and sleeping, and an almost total lack of private space. This is tempered somewhat in Western retreats which often cannot accommodate extreme collective living. Ch'an/Zen retreats embody the Japanese expression, "The nail that sticks out gets hammered in." A Japanese Zen retreat can be a pounding. Submission to coordinated group functioning is emphasized. The road to no-self is through group-self or so it might seem to non-Asians.

In contrast, the Forest Tradition while having group gatherings, usually for daily chanting and dhamma talks, is a more individual exercise. The institutional forms express a more underlying difference in temperament and style rather than doctrine. The image of the solitary sadhu wandering the forest of India lurks about the forest orders more than it does Ch'an/ Zen. The *kuti* or hut like residence of the monk provides the private space not available incold weather collectivist Zen. Although practice happens under the evaluative gaze of the abbot, mediation, walking, and sitting is often done in seclusion. A phrase such as "Up to you" (*lao dair*, Th.) in relation to scheduling your activity was never heard at a Zen temple or retreat center. One listened for the bells and clappers not your own rhythm. How long should one meditate? Zen centers prescribed set periods. Forest teachers suggested sitting and then after deciding to stop sitting sit longer. One felt one's way into the practice. I was lectured that wisdom lay not at the end of the path but had to be exercised all along the way.

The individualist leanings of the Forest Tradition carry over to an affinity for solitary practice. Monks were required to spend at least three rains retreats (three years) practicing under the daily direction of their seniors. After this seasoning period, monk would be allowed to do solitary retreats. This opportunity was generally well anticipated. Solo practice promoted in the Forest Tradition with monks practicing a distance away from their teacher even when traveling together. While of course Western Zen students are free to find private individual practice opportunities, as a regular feature of practice this is absent. So along with a more collectively regulated practice, Zen students were not actively encouraged to take up solitary practice. The hermits of China are a counterpoint.

The less formal group regime of the Forest Tradition should by no means be judged as lax. Group sitting periods might entail knee throbbing long rounds. All night meditations during rains retreat, but also voluntarily undertaken in personal practice, were encouraged. Students in both traditions will find the warrior mentality at work. Each style has its own challenges and practitioners need to judge where the greater benefit lay.

Surrendering to the collective and benefiting from its support may be as valuable as taking responsibility (and wisdom) for self-regulation.

Along with the individualist/collectivist distinction, the gradual/sudden enlightenment distinction is a second dichotomy that has practical implications. Sudden enlightenment is the signature feature of Zen. The Southern School of Ch'an asserted its superiority over its Northern rivals in large part on its sudden and presumably superior path to awakening. With the direct encounter with the Theravada, sudden enlightenment was again the badge of distinction. A long-term Zen Buddhist practitioner now residing in Thailand was considering attending a Theravada retreat. His Zen teacher warned him that this was a gradual path and therefore posed a contradiction to his prior training.

Aside from sectarian competition and territorial claims, the gradual/sudden contrast shapes the forest and mountain approaches to meditation. The initial stages of meditation are similar in seeking to lay a concentrative, samadhi, foundation. Both use an auditory object (Buddho/Mu) or the breath to calm and focus the mind. With the disciple's advance into the higher reaches of meditation the actual implications of sudden/gradual become apparent. The Forest Tradition insists on the application of samadhi power to the task of investigative insight primarily of impermanence and the body. Ch'an/Zen, on the other hand, focuses directly on the essence or nature of mind. The forest path proceeds by the examination of the conditioned world while Ch'an seeks an immediate breakthrough to the Unconditioned. The choice facing any would be practitioner is not who gets there sooner but the how one travels.

The mountain and forest traditions are not merely techniques and actions. The individualistic/collectivist styles and the sudden/gradual paths are embedded in a web of social relations. Here again are contrasts that need recognition. As already noted, the monk holds an august and more distant position toward the layperson in the Theravada tradition, despite the friendly relations between villagers and local monks. This carries on in the Forest Tradition even in the West where there is more care is taken in proper sitting positions and more restricted interaction with the opposite gender than in Ch'an/Zen centers. One Thai ordained Western monk (not in the Kammatthana proper) shared that he disrobed because he found the robe created too great a barrier between him and laypersons. Furthermore, Japanese Zen has attenuated the lay/monk distinction with the married priest. Much has been said about the veneration of the Zen teacher, but this did not inevitably extend to monks and nuns. This contrasts with the Theravada view of the monk as worthy of veneration and a pristine source of merit. The robe is an object of respect. The monk is an *ong*, a Thai grammatical classifier of the Buddha and his images.

An additional difference and consideration for both genders is that practice in the Forest Tradition means being largely in a male world. Certainly, laypersons and women play a role, but they are decidedly more marginal. Even in the West the Forest Tradition is male led in contrast to the prominence of women in the Zen tradition. As discussed, there is no Forest Monastery of men and women lead by a woman as in Western Zen. For all the ongoing concerns of latent sexism, Zen women have attained a level of prominence and power unmatched in the Kammatthana based schools. Beyond straight forward authority issues; this gives different hue to centers of the two schools. In conclusion, the key differences to be negotiated by a practitioner moving from one tradition to the other are their individualist/collectivist stances on the organization of practice and temple life, and their meditations reflecting gradual/sudden enlightenment. While teacher/student relations are markedly similar, the Forest Tradition is more monk-centered, meaning male and ordained.

A Final Note

The affinities between the forest and mountain schools have been investigated. A key cluster of characteristics, the soteriological project, self-cultivation, and charismatic leadership have been identified as interlocking drivers of commonalities. While the similarities are intriguing there is no suggestion that the two traditions are identical. The above noted contrasts demand intelligent navigation and not mere conflation by either researchers or Buddhists. Each school provides the other a platform for self-reflection which can only engender Buddhism more awake to its own contours and leanings. The recognition of similarities invites mutual appreciation of differences. They are not mutually exclusive. In the words of a Western psychologist, Fritz Perls, "Contact is the appreciation of differences." Hopefully Buddhism has entered into an epoch of more than mere inter-branch communication but of genuine, authentic contact.

References

McDaniel, J. (2011). *The lovelorn ghost and the magical monk.* New York: Columbia University Press.

Roebuck, V.J.(2010). *The Dhamadpada.* London: Penguin Books.

Sheng Yen. (2003). Silent illumination. *Buddhadharma,* (Fall), 16–20.

Walpola, R. (1978b). *Zen & the taming of the bull.* London: Gordon Frazier.

References

Amaro, B. (1983). *Silent rain*. Forest Sangha Publications.org. chapter 4

Clarke, R. (1973). *Faith in mind*. Buffalo, NY: White Pine.

Cleary, T. (1986). *Shobogenzo: Zen essays by Dogen*. Honolulu: University of Hawaii Press.

Cleary, T. (2001). *Rational Zen*. Boston, MA: Shambhala Press.

Deshimaru, T. (2012). *Mushotoku mind*. Kindle ebook edition.

Dhamma Wheel. (2009). Retrieved April 4, 2015, from http:\\www.dhammawheel.com/viewtopic.php?t=1000

Erickson, M. (1976). *Hypnotic realities*. New York: Irvington Publishers.

Erickson, M. (1996). *The wisdom of Milton Erickson* (Vol. 2). Bethel, CT.

Fuller, P. (1995). *The notion of ditthi in Theravada Buddhism*. New York: Routledge Curzon.

Gombrich, R. (2006). *Theravada Buddhism: From ancient Benares to modern Colombo*. New York: Routledge.

Hsu Yun. (1974). *Empty cloud: The autobiography of the Chinese Zen Master Hsu Yun* (C. Luk, Trans.). Rochester, NY: Empty Cloud Press.

Ireland, J. (1997). *The Udana and the Itivuttaka Sutta*. Kandy, Sri Lanka: Buddhist Publication Society.

Jorgensen, J. (2005). *Inventing Huineng*. Sinica Leidensia.

Liem, T. (2005). *No worries*. Ubon: Wat Pah Nanachat.

Maha Bua, B. (1995b). *Forest dhamma*. Udon Thani: Forest Dhamma Books.

Maha Bua, B. (1997). *Straight from the heart*. Udon Thani: Forest Dhamma Books.

Members of Nanachat. (n.d). *Forest path*. Ubon: Wat Pah Nanachat.

A.R. Lopez, *Buddhist Revivalist Movements*,
DOI 10.1057/978-1-137-54086-7

Nyanaponika. (1973). *The heart of Buddhist meditation*. York Beach, ME: Weiser Press.

Pasanno, A., & Amaro, A. (2012). *Achan Chah's teachings on nature*. Redwood Valley, CA: Abayagiri Monastery.

Schedneck, B. (2010). *Forest life stories 2*. Retrieved April 5, 2015, from http: // www.wanderingdhamma.org

Schedneck, B. (2015). *Thailand's international meditation centers*. New York: Routledge.

Sekida, K. (2005). *Zen training*. Boston, MA: Shambala.

Sekida, K., & Grimstone, A. V. (2005). *Two Zen classics*. Boston, MA: Shambhala.

Shaw, S. (2006). *Buddhist meditation an anthology of texts*. New York: Routledge.

Spiro, M. (1970). *Buddhism and society: A great tradition*. New York: Harper and Row.

Sujato, A. (n.d.). Retrieved August 9, 2015, from https://sujato.wordpress.com

Suzuki, D. T. (1972). *Zen doctrine of no-mind*. York Beach, ME: Weiser.

Suzuki, S. (1999). *Branching streams flow in the darkness*. Berkeley, CA: University of California Press.

Thanissaro, B. (2015a). *Noble strategy*. Valley Center, CA: Metta Forest Monastery.

Thanissaro, B. (2015b). No-self or not-self? In *Noble Strategy*. Valley Center, CA: Metta Forest Monastery.

Thich Nhat, H. (1974). *Zen keys*. Garden City, NY: Anchor Press.

Weber, M. (1993). *The sociology of religion*. Boston, MA: Beacon Press.

Wu, J. (2003). *The golden age of Zen*. Canada: World Wisdom.

Index

A.R. Lopez, *Buddhist Revivalist Movements*,
DOI 10.1057/978-1-137-54086-7

W

Y

Printed by Printforce, the Netherlands